9th National Seminar on Medical Physics

(NSMP 2014)

Journal of Physics: Conference Series Volume 546

Putrajaya, Malaysia
5 April 2014

ISBN: 978-1-63439-595-3
ISSN: 1742-6588

Printed from e-media with permission by:

Curran Associates, Inc.
57 Morehouse Lane
Red Hook, NY 12571

Some format issues inherent in the e-media version may also appear in this print version.

Printed by Curran Associates, Inc. (2015)

For permission requests, please contact the Institute of Physics
at the address below.

Institute of Physics
Dirac House, Temple Back
Bristol BS1 6BE UK

Phone: 44 1 17 929 7481
Fax: 44 1 17 920 0979

techtracking@iop.org

Additional copies of this publication are available from:

Curran Associates, Inc.
57 Morehouse Lane
Red Hook, NY 12571 USA
Phone: 845-758-0400
Fax: 845-758-2634
Email: curran@proceedings.com
Web: www.proceedings.com

TABLE OF CONTENTS

MEDICAL IMAGING

NUCLEAR MEDICINE

RADIATION THERAPY

RADIATION PROTECTION AND DOSIMETRY

BIOMEDICAL ENGINEERING

Journal of Physics: Conference Series **546** (2014) 011001 doi:10.1088/1742-6596/546/1/011001

Preface

The National Seminar on Medical Physics (NSMP) is a scientific conference organised every two years by the Malaysian Association of Medical Physics (MAMP). Its purpose is to provide a platform for researchers, medical physicists and clinicians from Malaysia and surrounding nations to discuss recent advances of research and development in medical imaging and radiotherapy.

NSMP 2014, the 9th national medical physics conference was held in Marriott Hotel, Putrajaya, Malaysia on 5 April 2014. The conference was organised in parallel to the College of Radiology (COR) Malaysia Scientific Meeting. The theme for the 9th NSMP is "Advances in Multidisciplinary Research and Clinical Practice".

About 65 participants from universities and hospitals participated in the conference. 17 oral contributions and 12 posters were presented at the conference. We had three invited lectures at the conference; two of the lectures were presented by international experts on state-of-the-art medical imaging and radiotherapy. The lectures were:

- "Hybrid imaging: research and clinical practice" by Prof David Townsend, A*STAR-National University Singapore Clinical Imaging Research Centre
- "Outline of treatment planning for carbon-ion radiotherapy" by Dr Nobuyuki Kanematsu, National Institute of Radiological Sciences, Japan
- "Implementing medical physics clinical training programme in Malaysia: challenges and experiences" by Dr Noriah Jamal, Malaysian Nuclear Agency

Many thanks to all invited speakers for their participation and to the Organising Committee members for all their hard work in making the conference happen. Thanks to all who submitted an abstract and making this a successful conference. The Scientific Committee members and reviewers are also thanked for reviewing the submitted manuscripts and improve the scientific quality of this proceedings. Finally, thanks to all who attended the conference and the sponsors for their financial support.

Conference Chairs
Wan A Kamil, Universiti Sains Malaysia (General Chair)
Rafidah Zainon, Universiti Sains Malaysia (Programme Chair)
Hafiz M Zin, Universiti Sains Malaysia (Scientific & Publication Chair)
Ahmad Taufek Abdul Rahman, Universiti Teknologi MARA (Publicity Chair)
Husaini Salleh, Malaysian Nuclear Agency (Finance Chair)

Conference Organising Committee
Mohd Haris Ridzuan Ooi Abdullah, Universiti Teknologi MARA
Azleen Mohd Zain, University Malaya Medical Centre
Abir Wardati Abd Latiff, Universiti Teknologi MARA
Siti Hajar Mohmad Salleh, Universiti Teknologi MARA

Journal of Physics: Conference Series **546** (2014) 011001 doi:10.1088/1742-6596/546/1/011001

Noramaliza Mohd Noor, Universiti Putra Malaysia
Nursyatina Abdul Raof, Universiti Sains Malaysia

Conference Scientific Committee and Manuscript Reviewers
International
David Townsend, A*STAR-NUS Clinical Imaging Research Centre, Singapore
Emma Harris, Institute of Cancer Research, UK
Matthew Blackledge, Institute of Cancer Research, UK
Naruhiro Matsufuji, National Institute of Radiological Sciences, Japan
Phil Evans, University of Surrey, UK
Supriyanto Ardjo Pawiro, University of Indonesia
Tan Jin Wooi, Brighton and Sussex University Hospitals NHS Trust, UK
Malaysia
Ahmad Nazlim Yusoff, Universiti Kebangsaan Malaysia
Ahmad Taufek Abdul Rahman, Universiti Teknologi MARA
Ashrani Aizzuddin Abd Rahni, Universiti Kebangsaan Malaysia
Chen Suk Chiang, Universiti Sains Malaysia
Hafiz M Zin, Universiti Sains Malaysia
Iskandar Shahrim Mustafa, Universiti Sains Malaysia
Jeannie Hsiu Ding Wong, University of Malaya
Mohd Hanafi Ali, Universiti Teknologi MARA
Mohd Zahri Abd Aziz, Universiti Sains Malaysia
Mohd Zubir Mat Johar, Universiti Sains Malaysia
Nahrizul Adib Kadri, University of Malaya
Ng Kwan Hoong, University of Malaya
Noramaliza Mohd Noor, Universiti Putra Malaysia
Rafidah Zainon, Universiti Sains Malaysia
Suhairul Hashim, Universiti Teknologi Malaysia
Ung Ngie Min, University of Malaya

Supporting Academic Organisation
Advanced Medical and Dental Institute, Universiti Sains Malaysia
College of Radiology, Academy of Medicine Malaysia

The proceedings consists of 22 manuscripts, organised into five different topics; medical imaging, nuclear medicine, radiation therapy, radiation protection and dosimetry, and biomedical engineering. All papers published in this volume of Journal of Physics: Conference Series have been peer reviewed through processes administered by the proceedings Editors. Reviews were conducted by expert referees to the professional and scientific standards expected of a proceedings journal published by IOP Publishing. We would like to thank all authors and reviewers for their contribution to this proceedings.

We look forward to seeing everyone in 2016 for the 10th anniversary.

Editors, Proceedings of NSMP 2014
Hafiz M Zin, Universiti Sains Malaysia
Ahmad Taufek Abdul Rahman, Universiti Teknologi MARA
Nahrizul Adib Kadri, University of Malaya
Rafidah Zainon, Universiti Sains Malaysia

Sponsorship Acknowledgements

The 2014 National Seminar on Medical Physics is sponsored by:

Abex Medical System Sdn Bhd
(authorised distributor of Elekta and Toshiba)

IDSMedical Systems (M) SdnBhd
(authorised distributor of GE)

Transmedic Pte Ltd
(authorised distributor of Accuray)

The Organising Committee is thankful to the above sponsors for their invaluable commitment and support.

Malaysian Association of Medical Physics
Advanced Medical & Dental Institute
Universiti Sains Malaysia
Bertam
13200 Kepala Batas
Pulau Pinang
Malaysia
Web: https://sites.google.com/site/mampwebsite/

Journal of Physics: Conference Series **546** (2014) 011002 doi:10.1088/1742-6596/546/1/011002

NSMP 2014 in images

Invited talks and opening ceremony

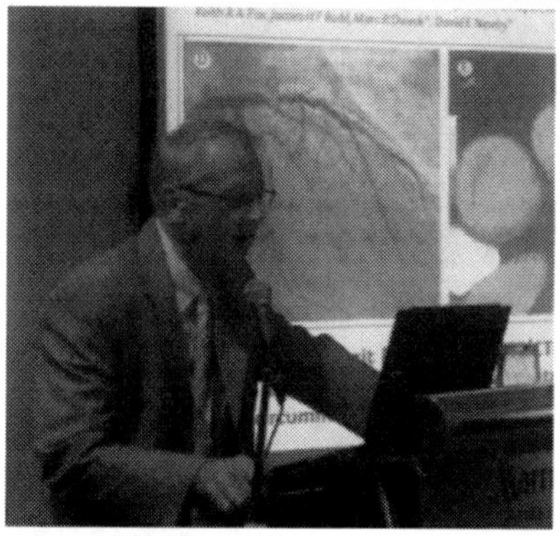

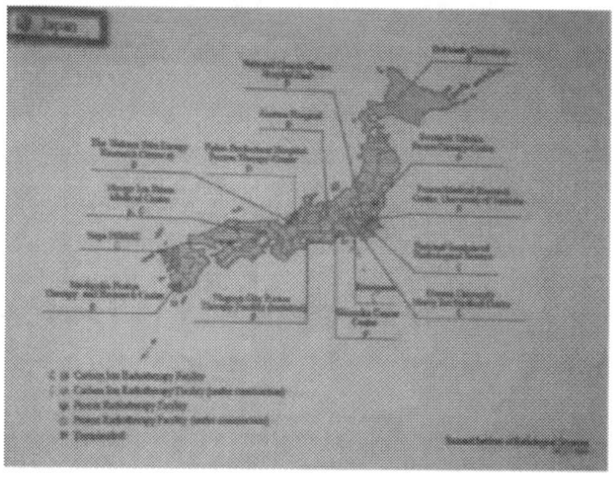

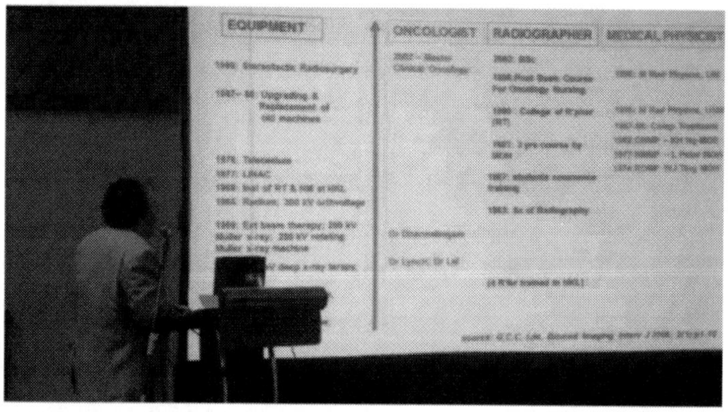

President of College of Radiology Malaysia, Dr Abdul Rahman (left) and President of Malaysian Association of Medical Physics, Assoc Prof Dr Wan A Kamil (right)

Oral presentations

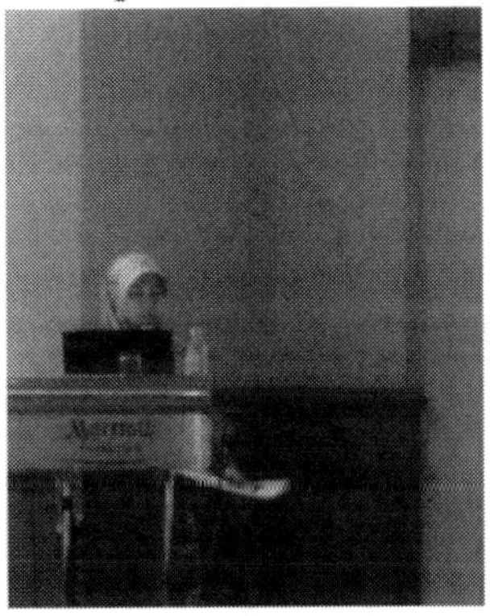

Journal of Physics: Conference Series **546** (2014) 011002 doi:10.1088/1742-6596/546/1/011002

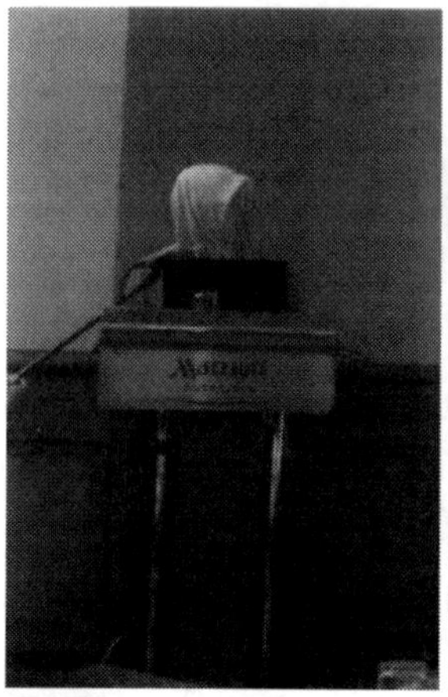

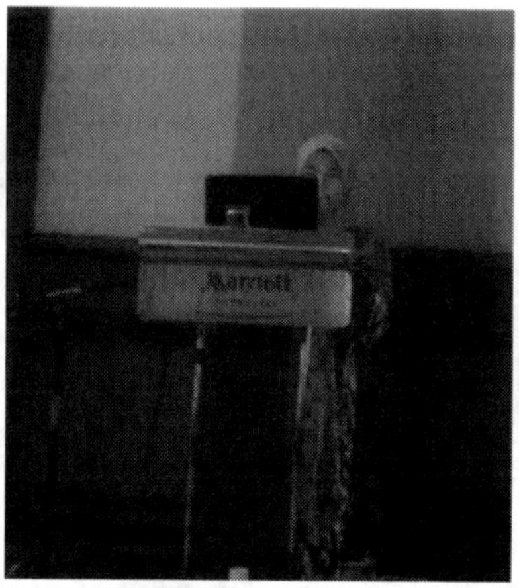

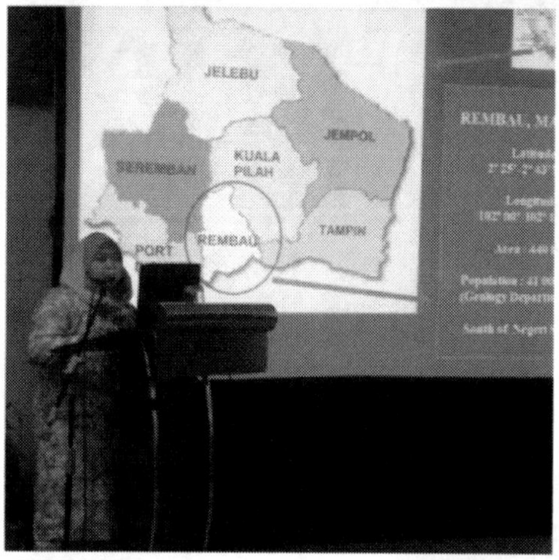

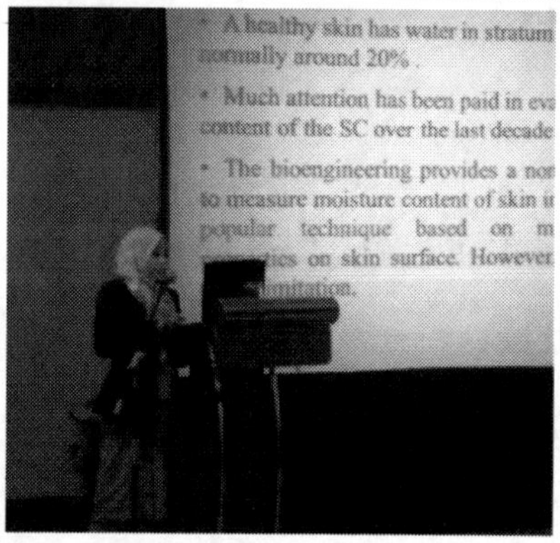

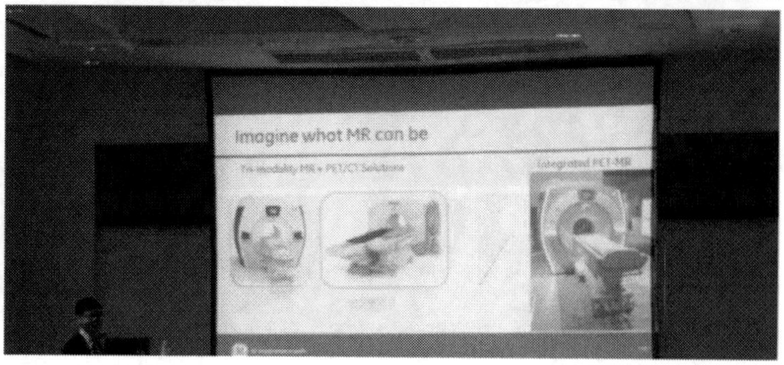

Journal of Physics: Conference Series **546** (2014) 011002 doi:10.1088/1742-6596/546/1/011002

Poster presentations

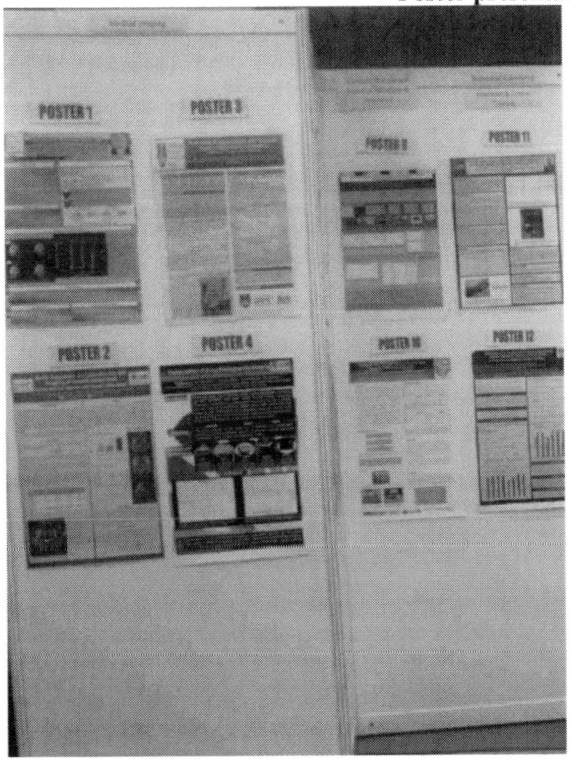

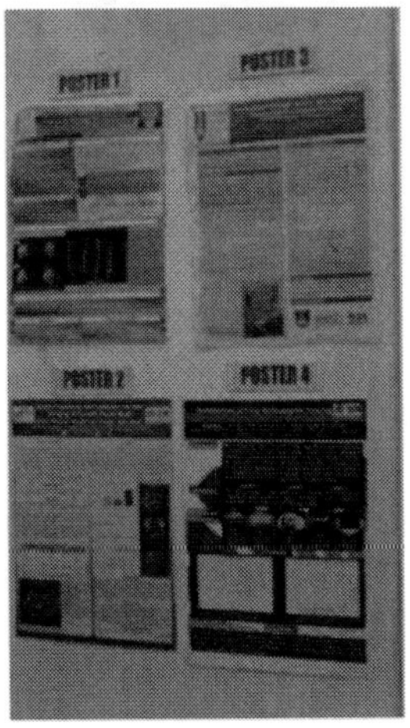

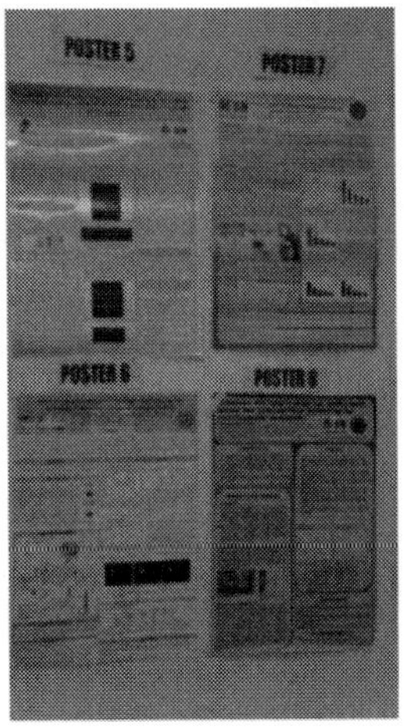

The delegates

Journal of Physics: Conference Series **546** (2014) 011002 doi:10.1088/1742-6596/546/1/011002

***Jumpa lagi di* NSMP 2016!**

Journal of Physics: Conference Series **546** (2014) 011003 doi:10.1088/1742-6596/546/1/011003

List of NSMP 2014 participants

No.	Name	Institution	E-mail
1	Abdul Rahman Mohamad	College of Radiology, Academy of Medicine of Malaysia	drrahman3033@gmail.com
2	Abir Wardati Abd Latiff	Faculty of Applied Science, Universiti Teknologi MARA Malaysia	abirwa2887@puncakalam.uitm.edu.my
3	Ahmad Taufek Abdul Rahman	Faculty of Applied Science, Universiti Teknologi MARA Malaysia	ahmadtaufek@salam.uitm.edu.my
4	Ahmad Yusairi Bani Hashim	Faculty of Manufacturing Engineering, Universiti Teknikal Malaysia Melaka	yusairi@utem.edu.my
5	Ahmed Almurshedi	Department of Physics, Faculty of Science, Universiti Teknologi Malaysia	fhahmed2@gmail.com
6	Anis Suhana Ahmad Sabri	Medical Physics Unit, University Malaya Medical Centre	suhana@ummc.edu.my
7	Ashrani Aizzuddin Abd. Rahni	Faculty of Engineering & Built Environment, Universiti Kebangsaan Malaysia	ashrani@eng.ukm.my
8	Azleen Mohd Zain	Medical Physics Unit, University Malaya Medical Centre	azleen@ummc.edu.my
9	Bakhtiar Mohd Ramadhan	Diagnostic Imaging Department, Hospital Serdang	bakhtiar.ramadhan@gmail.com
10	David Townsend	A*STAR-National University Singapore Clinical Imaging Research Centre, Singapore	david_townsend@circ.a-star.edu.sg
11	Elias Saion	Department Of Physics, Faculty Of Science, University Putra Malaysia	emansaion@gmail.com
12	Ernee Sazlinayati Othman	Pusat Asasi, Universiti Teknologi MARA, Puncak Alam	ernees2922@puncakalam.uitm.edu.my
13	Hafiz Mohd Zin	Advanced Medical and Dental Institute, Universiti Sains Malaysia	hafiz.zin@amdi.usm.edu.my
14	Husaini Salleh	Malaysian Nuclear Agency	husaini@nuclearmalaysia.gov.my
15	Jess Ooi	ABEX Medical System Sdn Bhd	jess_ooi@abexmedical.com.my
16	Laila Kalidah Junet	Kulliyyah of Allied Health Sciences, International Islamic University Malaysia	lailakalidahjunet@gmail.com
17	Lee Choong Hong	ABEX Medical System Sdn Bhd	chlee@abexmedical.com.my
18	Lim Chun Jek	GE Healthcare, ASEAN	chunjek.lim@ge.com
19	Lim Lih Chee	Radiotherapy Department, Lohguanlye Specialist Centre	lihchee@yahoo.com

20	Mastura Mohamad	School of Physics, Universiti Sains Malaysia	mastura.mohamad02@gmail.com
21	Mazlina Ahmad Fua'ad	KPJ Ipoh Specialist Hospital	mazlinafuaad@ish.kpjhealth.com.my
22	Md Saion Salikin	Faculty of Health Sciences, Universiti Teknologi MARA Malaysia	saionsalikin@gmail.com
23	Mohamad Aminudin Said	Institut Kanser Negara, Putrajaya	aminhpj@gmail.com
24	Mohd Fahmi Mohd Yusof	School of Physics, Universiti Sains Malaysia	mfahmi@usm.my
25	Mohd Hafizi Mahmud	Centre for Diagnostic Nuclear Imaging, Universiti Putra Malaysia	hafizimahmud@yahoo.com
26	Mohd Haris Ridzuan Ooi Abdullah	Faculty of Applied Science, Universiti Teknologi MARA Malaysia	harrisridzuan@gmail.com
27	Mohd Khairunizam Abdullah	Physics Unit, Radiotherapy & Oncology Department, Hospital Kuala Lumpur	knizam2105@gmail.com
28	Mohd Tarmizi Saidin	Physics Unit, Radiotherapy & Oncology Department, Hospital Kuala Lumpur	tarmizi7879@yahoo.co.uk
29	Mohd Zahri Abdul Aziz	Advanced Medical and Dental Institute, Universiti Sains Malaysia	zahri@amdi.usm.edu.my
30	Mohd Zulfadli Adenan	Department of Physics, Universiti Putra Malaysia	ibnu_soleh83@yahoo.com
31	Muhammad Faidzullah Mansor	Physics Unit, Radiotherapy & Oncology Department, Hospital Kuala Lumpur	muhammad_faidzullah@yahoo.com
32	Muhammad Nazmi Mazlan	Physics Unit, Radiotherapy & Oncology Department, Hospital Kuala Lumpur	nazmi_mazlan@yahoo.com
33	Muhammad Safwan Ahmad Fadzil	Faculty of Medicine and Health Sciences, Universiti Putra Malaysia	drsafwan_radiology@yahoo.com
34	Munira Mohd Rejab	Medical Physics Unit, University Malaya Medical Centre	munirarejab@ummc.edu.my
35	Nabilah Solehah Ahmad Tamrin	Faculty of Defence Science and Technology, Universiti Pertahanan Nasional Malaysia	nabilahsolehah@gmail.com
36	Najmeh Deyhimihaghighi	Department Of Physics, Faculty Of Science, University Putra Malaysia	parissacay@yahoo.com
37	Nazifah Abdullah	Kulliyyah of Allied Health Sciences, International Islamic University Malaysia	nazifah@unisza.edu.my
38	Nik Nur Hikmah Ramli	Faculty of Medicine and Health Sciences, Universiti Putra Malaysia	nnhikmah1305@yahoo.com
39	Nobuyuki Kanematsu	National Institute of Radiological Sciences, Japan	nkanemat@nirs.go.jp
40	Noor Arda Adrina Daud	Centre of Foundation Studies, Universiti Teknologi MARA	ardaadrina@hotmail.com
41	Nor Azwa Yunus	Faculty of Applied Science, Universiti Teknologi MARA Malaysia	norazwa0608@yahoo.com
42	Nor Eliana Norbani	Faculty Applied Sciences, Universiti Teknologi MARA Malaysia	norelianaorbani@yahoo.com
43	Noramaliza Mohd Noor	Department of Imaging, Universiti Putra Malaysia	noramaliza@upm.edu.my
44	Norhanna Sohaimi	Kull of Allied Health Sciences, International Islamic University Malaysia	norhanna82@yahoo.com
45	Noriah Jamal	Malaysian Nuclear Agency	noriahj@nuclearmalaysia.gov.my
46	Norzalina Zulkifli	Physics Unit, Radiotherapy & Oncology Department, Hospital Kuala Lumpur	zalina_zulkifli@yahoo.com

47	Nursyatina Abdul Raof	Advanced Medical and Dental Institute, Universiti Sains Malaysia	syatina@amdi.usm.edu.my
48	Nurul Arina Zahaimi	Faculty of Applied Science, Universiti Teknologi MARA Malaysia	rinazahaimi@yahoo.com
49	Nurul Hanim Nasaruddin	Faculty of Health Science, Universiti Kebangsaan Malaysia	hanimnasaruddin@gmail.com
50	Prisca Lee	Transmedic Group Sdn Bhd	prisca.lee@transmedicgroup.com
51	Rafidah Zainon	Advanced Medical and Dental Institute, Universiti Sains Malaysia	rafidahzainon@amdi.usm.edu.my
52	Safri Zainal Abidin	School of Health Sciences , Universiti Sains Malaysia	safri@usm.my safrivirgo@gmail.com
53	Siti Aishah Ibrahim	Department of Physics, Faculty of Science, Universiti Teknologi Malaysia	saishahibrahim90@gmail.com
54	Siti Hajar Mohmad Salleh	Applied Science Department, Universiti Teknologi MARA Pulau Pinang	sitiha2902@ppinang.uitm.edu.my
55	Siti Sarah Mat Isa	Kulliyyah of Allied Health Sciences, International Islamic University Malaysia	ssmi1206@gmail.com
56	Siti Shuwaibah Che Omar	Department of Physics, Faculty of Science, Universiti Teknologi Malaysia	sitishuwaibah@yahoo.com.
57	Siti Suhana Ismail	Faculty of Health Science, Universiti Kebangsaan Malaysia	ctsuhana87@gmail.com
58	Sity Noor Ayseah Dzulkafli	Faculty of Health Sciences, Universiti Teknologi MARA Malaysia	ayseah@gmail.com
59	Sofina Tamam	School of Medical Sciences, Universiti Sains Malaysia	sofina_tamam@yahoo.com
60	Suzana Mat Isa	Advanced Medical and Dental Institute, Universiti Sains Malaysia	suzana@amdi.usm.edu.my
61	Wan Ahmad Kamil Wan Abdullah	Advanced Medical and Dental Institute, Universiti Sains Malaysia	wkamil@amdi.usm.edu.my
62	Wan Mohd Haiziliy Wan Hassan	Medical Physics Unit, University Malaya Medical Centre	haizily@ummc.edu.my
63	Wan Nordiana W Abd Rahman	School of Health Sciences , Universiti Sains Malaysia	wandiana@usm.my
64	Zulaikha Jamalludin	Medical Physics Unit, University Malaya Medical Centre	ikha@ummc.edu.my

Peer review statement

All papers published in this volume of *Journal of Physics: Conference Series* have been peer reviewed through processes administered by the proceedings Editors. Reviews were conducted by expert referees to the professional and scientific standards expected of a proceedings journal published by IOP Publishing.

Quantication and analysis of respiratory motion from 4D MRI

Ashrani Aizzuddin Abd. Rahni[1], Emma Lewis[2] and Kevin Wells[2]

[1] Department of Electrical, Electronic and System Engineering, Faculty of Engineering and Built Environment, Universiti Kebangsaan Malaysia, 43600 UKM Bangi, Selangor, Malaysia

[2] Centre for Vision, Speech and Signal Processing, University of Surrey, Guildford, Surrey, GU2 7XH, United Kingdom

E-mail: ashrani@eng.ukm.my

Abstract. It is well known that respiratory motion affects image acquisition and also external beam radiotherapy (EBRT) treatment planning and delivery. However often the existing approaches for respiratory motion management are based on a generic view of respiratory motion such as the general movement of organ, tissue or fiducials. This paper thus aims to present a more in depth analysis of respiratory motion based on 4D MRI for further integration into motion correction in image acquisition or image based EBRT. Internal and external motion was first analysed separately, on a per-organ basis for internal motion. Principal component analysis (PCA) was then performed on the internal and external motion vectors separately and the relationship between the two PCA spaces was analysed. The motion extracted from 4D MRI on general was found to be consistent with what has been reported in literature.

1. Introduction

It is well known that respiratory motion affects image acquisition and also external beam radiotherapy (EBRT) treatment planning and delivery [1]. Many approaches to manage respiratory motion are also known and researched. The method of motion compensation or corrections depends on the assumption of the corrupting motion. Often the existing approaches are based on a generic view of respiratory motion such as the general movement of organ, tissue or fiducials [2]. More recently there has been interest in using dynamic volumetric imaging such as 4D MRI for more accurate quantification of motion. There have also been recent studies aiming to generate 4D CT from 4D MRI with the indented benefit of less ionising radiation and capability of capturing variability in motion [3, 4]. However, analysis based on 4D MRI may still describe respiratory motion in a rather generic way [5]. This paper builds on previous work [6, 7], aiming to present a more in depth analysis of respiratory motion based on 4D MRI for further integration into motion correction in image acquisition or image based EBRT.

2. Methodology

An open access 4D MRI dataset [8] was used to obtain respiratory motion for five volunteers. As described in [8], the data was acquired using a 1.5 T Philips Achieva™ scanner with a 32-channel coil. The temporal resolution was 0.7 s per volume, with each volume having a dimension of $336 \times 45 \times 336$ voxels and a voxel size of $1.48214 \times 5.5 \times 1.48214$ mm^3. For this paper, only the motion for the first volunteer was analysed as a proof of concept of the methodology.

For a more detailed analysis of the motion to be found, the major organs were delineated using a semi-automated segmentation procedure as in [6]. The segmented organs are the lungs, heart, liver, spleen, stomach and kidneys. The major airways i.e. trachea and primary bronchi, was also segmented. The respiratory system itself was segmented using a more automated procedure as described in [9]. The resulting segmentation is shown in Figure 1 using ITK-SNAP.

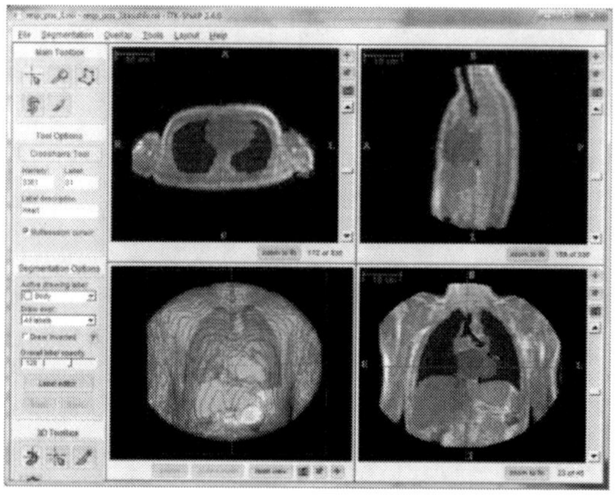

Figure 1. Display of slices in three orthogonal planes with a rendering of the segmented anatomy (lower left) in ITK-SNAP. The rendered anatomy are the lungs (blue), heart (red), liver (magenta), stomach (peach), spleen (cyan), kidneys (dark red), major airways (dark blue) as well as the body (yellow).

Respiratory motion was extracted using a free form deformation based image registration i.e. the NiftyReg registration package [10, 11]. The default NiftyReg parameter values were used and sum of square differences was used as the registration basis. The motion was calculated for each voxel of the image as well as averaged values found per organ. Motion was quantified as a vector in the image coordinate system (superior-inferior (SI), anterior-posterior (AP), left-right (LR), referred to hereafter as the z-, y- and x-axes) as well as in terms of magnitude of the motion vector.

Additionally, motion of the anterior surface of the body was also found using the procedure described in [7]. This ensures that the motion can be found to sub-voxel accuracy as well as using a different method than the one used for extracting internal motion. The purpose of detecting surface motion serves as a simulation of motion detected from a markerless camera surrogate, such as the VisionRT system. The correspondence between internal and external motion is also analysed as was performed in [7].

3. Analysis and results

In this section, the internal and external motion was analysed separately as well as the correlation between them in subsections 3.1 and 3.2 respectively.

3.1. Internal and External Motion

As in [6], internal motion was quantified in the x-, y- and z-axes on a per-organ basis as well as over all major organs. Additionally the average magnitude was found over all organs. This is shown in Figures 2(a) and (b). The error bars signify the standard deviation.

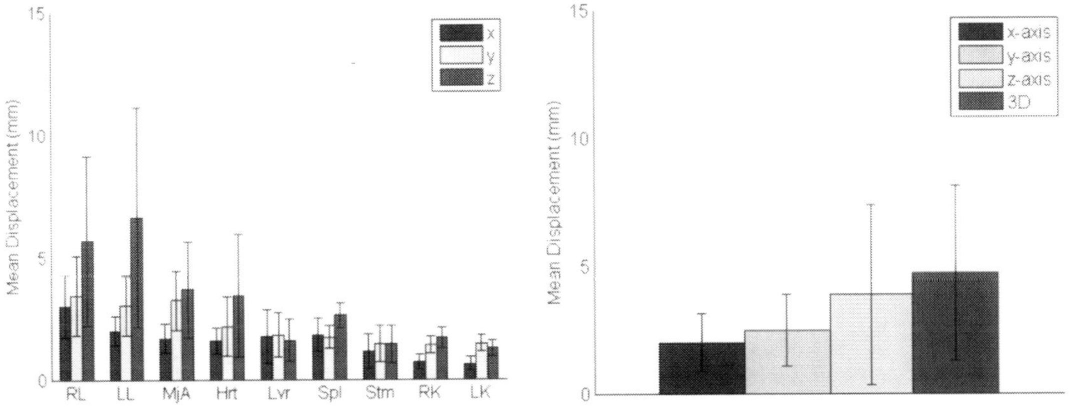

(a) Maximum component displacement for each organ.

(b) Maximum displacement over all organs.

Figure 2. Analysis of the mean trajectory of all voxels in the chosen organs. The labels of the chosen organs are: RL: right lung, LL: left lung, MjA: major airways, Hrt: heart, Lvr: liver, Stm: Stomach, Spl: spleen, RK: right kidney and LK: left kidney.

From Figure 2(a), it can be seen that in general, the proportion of motion between the three orthogonal directions are as expected, i.e. larger in the z-axis (SI motion). Additionally the average magnitude of motion over all organs as shown in Figure 2(b) is of a reasonable value i.e. just above 5 mm.

As in [6], the internal motion was compared with a previous study [12], where organ motion was found via manual segmentation on 20 organs. The motion found from this study is as shown in Figure 3(a).

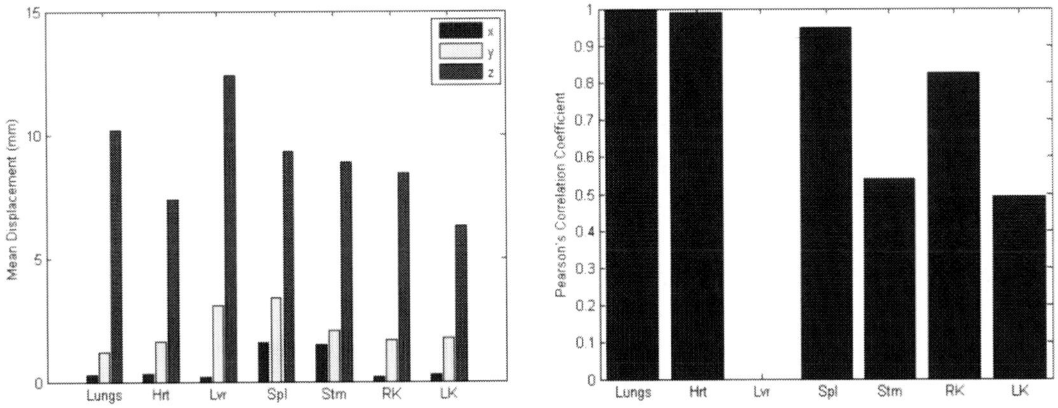

(a) Mean displacement from a previous study.

(b) Correlation of component displacements.

Figure 3. (a) Mean displacement from a previous study [12]. The labels for the other organs with the exception of the lungs are the same as Fig. 2. (b) Correlation of component displacements.

In Figure 3(a), only the mean value is available as well as a single value for the lungs. The z-axis motion of the lungs is also not available and hence it is assumed to be the average of the z-axis motion of the organs immediately below the diaphragm (liver, stomach and spleen).

The correlation between the results of this previous study and that shown in Figure 2(a) was also found. The correlation was quantified over the proportion of motion over the x-, y- and z-axes. This

Journal of Physics: Conference Series **546** (2014) 012001 doi:10.1088/1742-6596/546/1/012001

measure of correlation is shown in Figure 3(b) showing a generally high correlation for all organs except the liver (on average 0.8 excluding the liver). This may be attributed to poor delineation due to the low resolution of the images. Note here that the magnitude of motion is not considered as it can differ greatly from person to person.

Lastly as a qualitative assessment, Figure 4(a) shows the motion detected on the anterior surface of the body (at a few selected points arranged in a 5 × 9 point grid) while Figure 4(b) shows the trajectory of a few selected points (in any of the orthogonal axes) of the major organs.

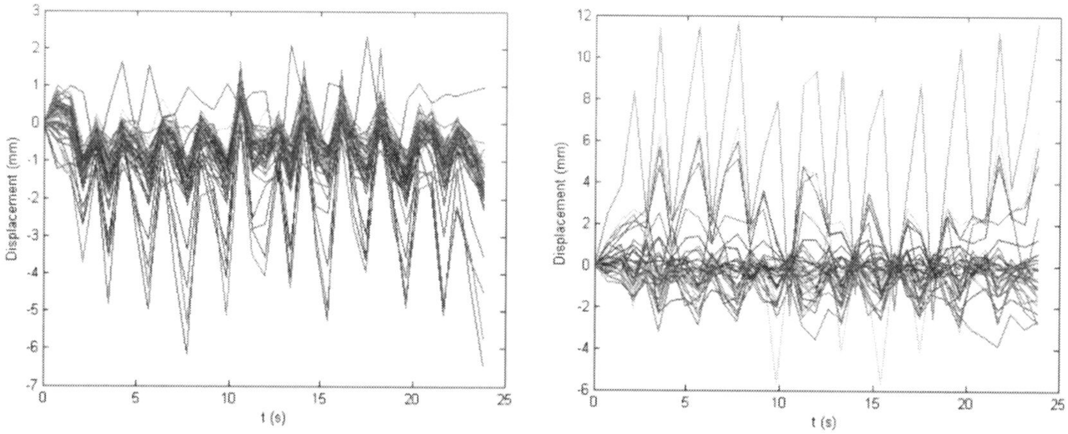

(a) Trajectory of selected body surface points. (b) Trajectory of selected organ points.

Figure 4. 1D trajectory of selected points of (a) the anterior body surface and (b) major organs.

From both Figures 4(a) and (b) it can be seen that the peaks occur at roughly the same time points and hence it can be said generally that there is correlation between the internal and external motion extracted using different methods. However, a more quantitative assessment is described in subsection 3.2.

3.2. Correspondence of Internal and External Motion

As the motion found is high dimensional (45 for external motion whereas internal motion is found over around 500,000 voxels), dimensionality reduction is applied to both internal and external motion separately using principal component analysis (PCA). The number of dimensions was reduced to one less than the number of data points i.e. 34, due to the rank of the data matrices for the internal and external motion respectively. Both types of motion in their respective PCA spaces are as shown in Figures 5(a) and (b).

To show the partial correspondence of internal and external motion, the correlation of the first PC of internal motion with all selected external points was found. This is shown in Figure 6(a) where the correlation is high (magnitude around ±1) except for a few points due to uncertainty from the low image resolution.

Additionally, the full correspondence was found by calculating the correlation of all PCs of internal motion with all PCs of external motion. This is shown in Figure 6(b) where only the first few respective PCs of internal and external motion show high correlation.

Lastly the first PC of internal motion is plotted against the first PC of external motion as shown in Figure 7. Here the phenomenon of hysteresis [13] is evident and serves as an additional reinforcement that the motion found is reasonable and can represent the true internal and external motion.

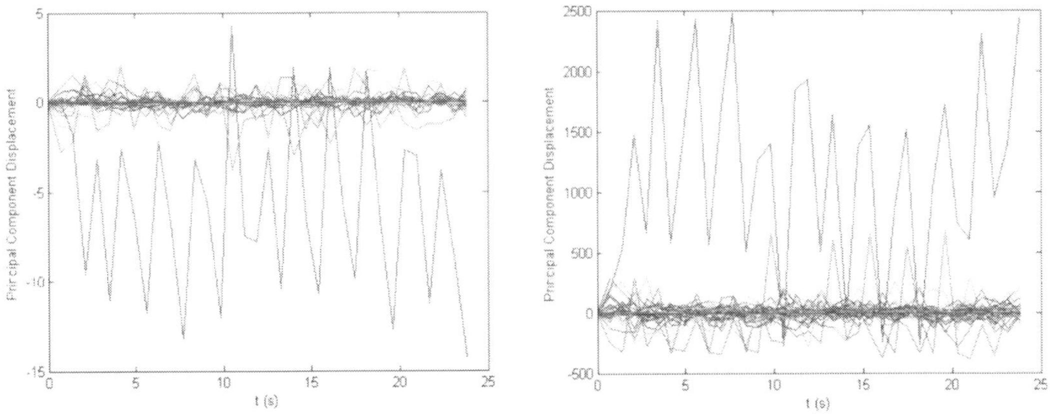

(a) Motion of the anterior body surface. (b) Motion of major organs.

Figure 5. Motion of (a) the anterior body surface and (b) major organs in their respective PCA spaces.

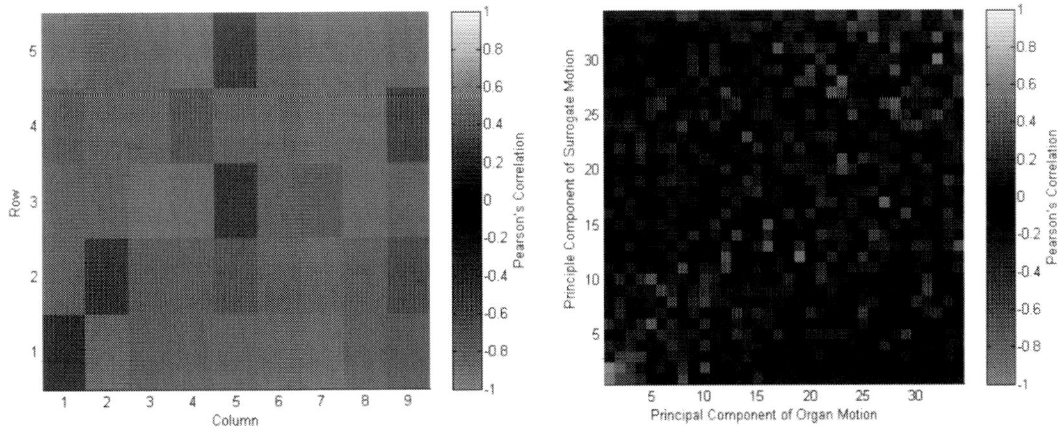

(a) Correlation of surface points. (b) Correlation between PCs of organ and surrogate motion.

Figure 6. Fig. (a) shows the correlation between the surrogates at selected surface points and the first PC of organ motion. Columns and rows refer to the grid of surface points. Fig. (b) shows the correlation between PCs of organ and surrogate motion.

4. Conclusion

The analysis in Section 3 shows that respiratory motion (both of the organs and of the anterior body surface) can be found from 4D MRI despite the low spatial and temporal resolution. This is strengthen by the results that show that the detected motion exhibits features reported in previous measurement. An advantage of the image based approach allows motion to be extracted over the whole field of view as opposed to motion found from fiducials.

However, there are still avenues for improvement in future work. Firstly the analysis can be extended to all five volunteers in the dataset as well as using larger datasets. Secondly, surface motion can be captured using actual cameras so as to become a totally separate source of measurement from internal motion which is found from dynamic volumetric images. Lastly a feasible motion compensation approach is to be implemented. While many approaches have been researched, there is still the question of an implementation which is clinically feasible [14].

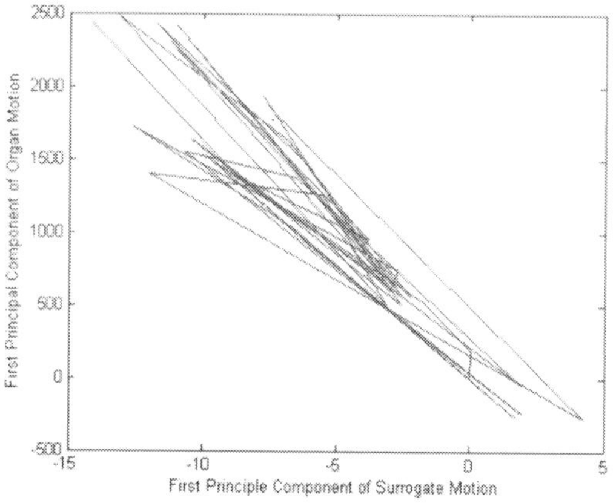

Figure 7. Joint trajectory for the first respective PCA projection of (internal) organ motion, and (external) surrogate motion.

Acknoledgement
Acknowledgement to Ehsan Golkar for help in organ segmentation.

References

[1] A reference Keall P J, Mageras G S, Balter J M, Emery R S, Forster K M, Jiang S B, Kapatoes J M, Low D A, Murphy M J, Murray B R, Ramsey C R, van Herk M B, Vedam S S, Wong J W and Yorke E 2006 The management of respiratory motion in radiation oncology report of AAPM Task Group 76 Tech. rep. American Association of Physicists in Medicine
[2] Yang W, Fraass B A, Robert R, Nissen N, Lo S, Jamil L H, Gupta K, Sandler H and Tuli R 2014 *Radiation Oncology* **9**
[3] Marx M, Ehrhardt J, Werner R and Handels H 2014 *Int J CARS* **9** 401 - 409
[4] Boye D, Lomax T and Knopf A 2013 *Med Phys* **40**
[5] Miquel M E, Blackall J M, Uribe S, Hawkes D J and Schaeffter T 2014 *Physica Medica* **29** 214 - 220
[6] Abd Rahni A A, Lewis E and Wells K 2013 *SPIE Medical Imaging: Image Processing* vol 8669 p 866930
[7] Abd Rahni A A, Lewis E and Wells K 2013 *SPIE Medical Imaging: Image Processing* vol 8669 p 86692Z
[8] Tsoumpas C, Buerger C, King A P, Mollet P, Keereman V, Vandenberghe S, Schulz V, Schleyer P, Schaeffter T and Marsden P K 2011 *Phys Med Biol* **56** 6597 - 6613
[9] Abd Rahni A A, Lewis E and Wells K 2013 *IEEE ICSIPA* pp 232 - 236
[10] Modat M, Ridgway G R, Taylor Z A, Lehmann M, Barnes J, Hawkes D J, Fox N C and Ourselin S 2010 *Comput. Methods Programs Biomed.* **98** 278 - 284
[11] Ourselin S, Stefanescu R and Pennec X 2002 *MICCAI* vol 2489 pp 140 - 147
[12] Segars W P, Mori S, Chen G T Y and Tsui B M W 2007 *IEEE NSS/MIC Conf Rec* vol 4 pp 2677 - 2679
[13] Escolar J and Escolar A 2004 *Histol Hispathol* **19** 159 - 166
[14] McClelland J R, Hawkes D J, Schaeffter T and King A P 2013 *Med Imag Anal* **17** 19 - 42

The effect of compensating filter on image quality in lateral projection of thoraco lumbar radiography

N A A Daud[1], M H Ali[2], N A Ahmad Nazri[1], N J Hamzah[1] and N A Awang[2]

[1]Centre of Foundation Studies, Universiti Teknologi MARA, Puncak Alam, Selangor, 42000 Malaysia
[2]Faculty of Health Sciences, Universiti Teknologi MARA, Puncak Alam, Selangor, 42000 Malaysia

E-mail: ardaadrina@hotmail.com

Abstract. The aim of this project was to study the effect of compensating filter on image quality in lateral projection of thoraco lumbar radiography. The specific objectives of this study were to verify the relationship between density, contrast and noise of lateral thoraco lumbar radiography using various thickness of compensating filter and to determine the appropriate filter thickness with the thoraco lumbar density. The study was performed by an X-ray unit exposed to the body phantom where different thicknesses of aluminium were used as compensating filter. The radiographs were processed by CR reader and being imported to KPACS software to analyze the pixel depth value, contrast and noise. Result shows different thickness of aluminium compensating filter improved the image quality of lateral projection thoraco lumbar radiography. The compensating filter of 8.2 mm was considered as the optimal filter to compensate the thoraco lumbar junction (T12-L1), 1 mm to compensate lumbar region and 5.9 mm to compensate thorax region. The addition of aluminium compensating filter is advantageous in terms of efficiency which saving radiograph film, workload of the radiographer and radiation dose to patient.

1. Introduction

Imaging body parts such as lateral thoraco lumbar radiography is a challenge due to different composition and density of thorax and lumbar area. The difference may cause non uniformity of optical density on radiograph. There are several techniques to overcome the limitation in lateral thoraco lumbar radiography such as using a high tube voltage technique, two exposure techniques and the use of compensating filter.

One advantage of using high kV technique is shorter exposure time [1]. The higher the energy, the more photons will penetrate the body, the less that are absorbed. Although this reduces patient dose, image quality is compromised due to reduced contrast of a preferentially forward scattered radiation [2], [3]. Another technique can be used to overcome this issue is known as two exposure technique where the images may be taken twice with different exposure parameters. It produces two images in two exposures; one is thoracic and the other is lumbar. However it has disadvantage of doubling the dose to the patient since the exposure is taken twice. Additionally, it increases the time and workload of the radiographer.

Filters are important for the purpose of patient protection and improving radiographic image quality [4]. It is used in medical imaging to attenuate and hardened the x-ray beam spectrum [5]. Increasing the thickness of the filter removes a great number of low energy photons that is absorbed in the body, thus reducing the radiation dose to patient [6].The usage of compensating filter is effective as it is designed specifically for anatomical area that consist of varying thickness and density [7]. One advantage of compensating filter is anatomy of significant varying thickness can be imaged with a single exposure. Other advantages are quick and ease of use [8].In this study, aluminium was used as compensating filter as it reduces the proportion of low energy photons which is absorbed in the body. It is used due to its low atomic number thus a suitable filter material for the purpose of tissue compensation [9].

The research on the use of compensating filter in radiography has been conducted by several authors [10], [11]. A special aluminium wedge filter has been used in scoliosis radiography. The thickness of the filter is 19 mm in the chest region and 6.5 mm in the abdomen region [10].Wieder and Adam use the trough filter made of aluminium for posteroanterior chest radiographs. It consists of a square plate of aluminium with a centre trough that helps to produce high quality chest radiograph [11].

In image quality, radiographic density is the measure of overall darkening of the image. Density is a logarithmic unit that describes the ratio between light incident on the film and light being transmitted through the film [12]. The higher the radiographic density shows the more opaque areas of the film. The lower the density represents more transparent areas of the film. In term of pixel depth, the values varies from 0 (intensity for black pixel) to 255 (intensity for white pixel). In X-ray radiography, noise is the relevant comparison to contrast. In this study, noise properties of an image are described in terms of standard deviation measured within a region of interest (ROI) [13].

The main purpose of this research is to study the effect of compensating filter on image quality in lateral projection of thoraco lumbar radiography. The first specific objective of this study is to verify the relationship between density, contrast and noise of lateral thoraco lumbar radiography using various thickness of compensating filter. Besides, the study is conducted to determine the appropriate filter thickness with the thoraco lumbar density.

2. Material and method

2.1. Radiography

The experiment was conducted in X-ray laboratory of UiTM Puncak Alam. The study was performed using X-ray unit model Philips with equivalent filtration of 2 mm aluminium. The source to image distance (SID) was set to 100 cm.

2.2. Phantom

A Whole Body Phantom PBU-50 was placed in lateral projection. It is full body anthropomorphic phantom consists skeleton, lungs, liver, mediastinum and kidneys embedded in KYOTOKAGAKU original soft tissue. The movable joints allow the positioning of lateral projection of thoraco lumbar radiography.

2.3. Compensating filter

The compensating filter was constructed from a single piece of aluminium of 10 cm x 10 cm dimensions. The aluminium thickness ranged between 1 mm to 11 mm. The filter is placed in front of the light beam diaphragm with the aid of a filter holder. A number of images were acquired with different thicknesses of filter, ranging from 1 mm to 11 mm, in steps of 1 mm.

2.4. Exposure factor

A series of AEC exposure was made to determine the optimum kVp and mAs without the aluminium compensating filter. Another series of exposure then was made to determine the optimum kVp and mAs with the aluminium compensating filter. In this experiment the optimum kVp and mAs was 85 kVp and 20 mAs. The exposures parameters were kept constant throughout the experiment while the compensating filter was inserted from least to the greatest so that the effect of increasing the thickness of aluminium compensating filter on image quality can be evaluated.

2.5. Image analysis
The images were processed and read by the REGIUS CR reader MODEL 210. The radiographs were imported to the KPACS software for the analysis purpose. The KPACS software was used to analyse the pixel value, contrast and noise of the images. The relationship between densities of thorax and lumbar with various thickness of compensating filter was verified. Besides, the appropriate filter thicknesses with thoraco lumbar density was determined.

Using (1) the Optical Density (OD) was calculated

$$Optical\ Density\ (OD) = 2 - \left[\frac{220 - Pixel\ Depth\ Value}{100}\right] \quad (1)$$

The additive factor for each region of thorax and lumbar were calculated by equation (2)

$$Additive\ factor = \frac{Pixel\ Depth\ value\ for\ thickness\ x}{Pixel\ Depth\ value\ with\ zero\ thickness} \qquad x = 1mm, 2mm \ldots .11mm \quad (2)$$

Thickness of Aluminium compensating filter required to compensate the thorax and lumbar region is the value which the ratio of optical density between thorax and lumbar equal to 1.It was calculated using equation (3)

$$Appropriate\ filter\ thickness = \frac{OD\ Thorax}{OD\ Lumbar} = 1 \quad (3)$$

3. Result and discussion

3.1. Pixel depth value analysis
The graph of pixel depth value versus thickness of aluminium filters for a step of T8, T9, T10, T11, T12 in thorax region and L1, L2 and L3 in the lumbar region was shown by the graph in Figure 1. As the filter thickness increased, the value of pixel slightly increased (density reduced). There was sudden change of pixel values between 5 to 6 mm of aluminium thickness due to the characteristics of x-ray. At this point, the compensation was effective as the filter compensate more than others.

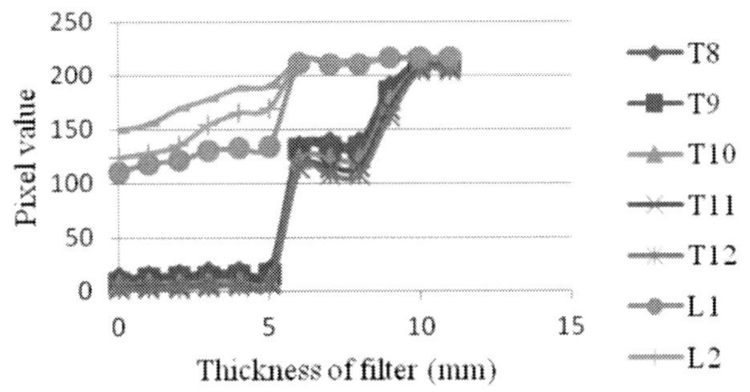

Figure 1. Graph of Pixel value vs Thickness of filter for different region of Thorax and Lumbar.

The compensating filter absorbs radiation according to the thickness of the filter. Without compensating filter, pixel depth value of lumbar L1-L3 is higher (low density) compared to the pixel depth value of thorax T8-T12 (high density) means that the thorax region was too dark (black) while the lumbar region shows the good quality image and acceptable in image diagnosis. More filter thickness is required to compensate the thorax region to decrease its density while maintaining the density of the lumbar region. However, increasing too much thickness of the filter improved the image of thorax but reduced the image quality of lumbar. Due to this reason we need to design one compensating filter with appropriate thickness in order to compensate both thorax and lumbar region so that the thoraco lumbar region can be imaged in one exposure.

3.2. Noise analysis

Figure 2 shows the graph of standard deviation in different regions when using different thicknesses of aluminium filter. The standard deviation represents the noise of an image. In relation to the contrast, noise increased as the contrast increased. The lower density difference means there is lower contrast. Contrast within a film increases with increasing density difference. In thorax region, the noise level slightly increased with increasing of filter thickness and recommended the maximum thickness is 6 mm. Thickness of 7 mm to 11 mm has reduced the image quality of the thorax region. In a lumbar region, noise decreased as the thickness of filter increased.

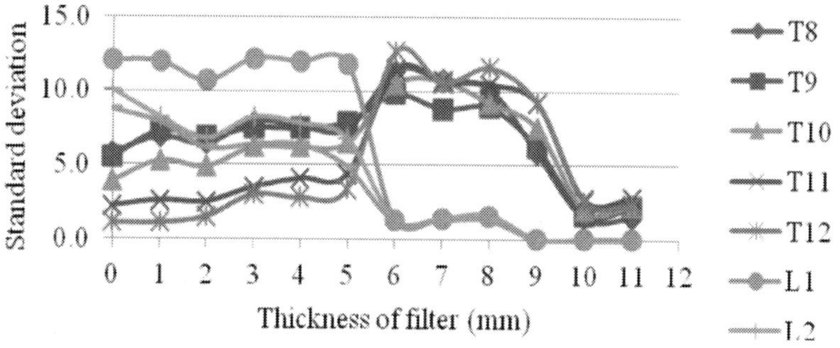

Figure 2. Graph of Standard deviation vs Aluminium Thickness for different region of Thorax and Lumbar.

3.3. Additive factor

In thorax region, the additive factor of pixel value was greater than 1 with the increasing of filter thickness. The density was highly reduced in the thorax region. The factor was approximately 1 for lumbar region since the density was almost high at lumbar region even there was no compensating filter involved.

Based on the additive factor calculation, the appropriate thickness of filter was 8 mm maximum in the thorax and 1 mm maximum in the lumbar region. Even though thickness of 6 mm, 7 mm and 8 mm contribute to the optimum density of thorax, thickness of 8 mm was preferable in term of dose in which it reduced more dose compared to the other two.

3.4. Design of wedge compensating filter

Figure 3 shows the graph of optical density vs thickness of filter in thorax and lumbar region respectively. In order to design the wedge compensating filter, the appropriate thickness of the filter was determined specifically for lumbar and thorax region by calculating equation (3).

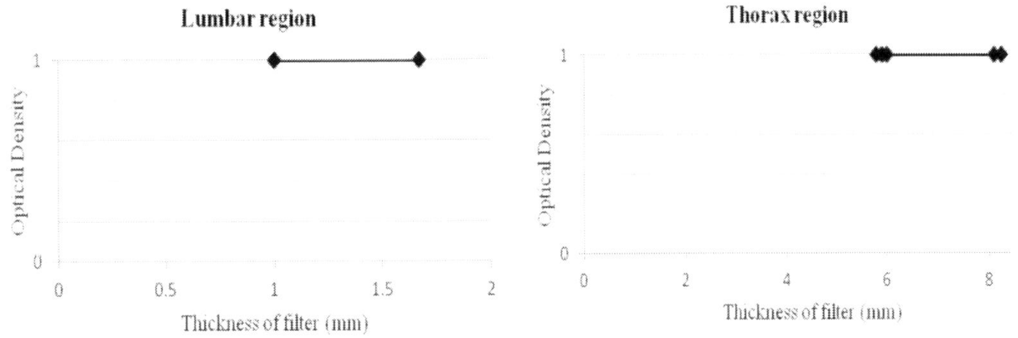

Figure 3. Graph of optical density vs thickness of aluminium filter in lumbar and thorax region

The shape of aluminum wedge compensating filter is shown in Figure 4.The wedge compensating filter was attached to the perspex holder and can be inserted into the x-ray collimator. The size of the filter was 10 cm x 10 cm. The minimum filter thickness was 1 mm in the lumbar region while 5.9 mm in the thorax region. The thickness increased towards the thoraco lumbar area contributes the maximum thickness of 8.2 mm in the region between thorax and lumbar region (thoraco lumbar junction) which is between T12-L1.

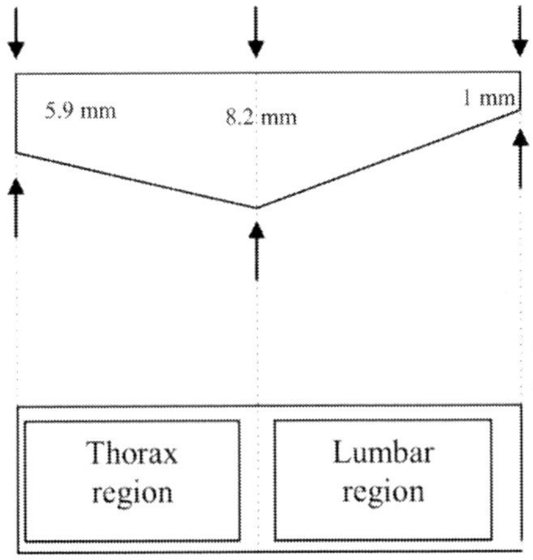

Figure 4. The design of wedge compensating filter for lateral projection of thoraco lumbar radiography

4. Conclusion

There are many advantages when using compensating filter for lateral projection of thoraco lumbar radiography. It was efficient in imaging the part of varying tissue thickness and density. It also improves the image quality of the radiography. The radiographer in this case has potentially saved the patient from being exposed to a large radiation because the design of compensating filter allows the single exposure instead of two. Therefore, saving the time and workload of the radiographer.

Acknowledgment
I would like to acknowledge Research Management Institute (RMI) Universiti Teknologi Mara for the Excellence Fund RIF (Research Intensive Faculty) grant no. 600-RMI/DANA 5/3/RIF (179/2012)

References
[1] Vassileva J 2004 A phantom approach to find the optimal technical parameters for plain chest radiography *Br J Radiology* **77** 648
[2] Martin C J 2007 The importance of radiation quality for optimisation in radiology *Biomed Imaging Interv J* **3**
[3] Almen A,Tingberg A,Mattsson S,Besjakov J,Kheddache S,Lanhede B,Mansson L G and Zankl M 2000 The influence of different technique factors on image quality of lumbar spine radiographs as evaluated by established CEC image criteria *Br J Radiology* **73** 1192
[4] Shockley V E,Kathren R L and Thomas E M 2008 Reconstruction of doses from occupationally related medical x-ray examinations *Health Phys* **95** 107
[5] Uffmann M and Schaefer-Prokop C 2009 Digital radiography:The balance between image quality and required radiation dose *European Journal of Radiology* **72** 202
[6] Goncalves A,Rollo J,Goncalves M,Haiter Neto F and Boscolo F 2004 Effects of Aluminum-Copper Alloy filtration on photon spectra,air kerma rate and image contrast *Braz Dent J* **15** 214
[7] Davidson R A 2001 Determination of radiographic characteristic of tissue compensation filters using a Compton scatter technique *Australas. Phys. Eng Sci.Med* **24**
[8] Alcaraz M and Garcia-Vera M 2009 Collimator with filtration compensator:clinical adaptation to meet European Union recommendation 4F on radiological protection for dental radiography *Dentomaxillofacial Radiology* **38** 413
[9] Martin C J 2007 Optimisation in general radiography *Biomed Imaging Interv J* **3**
[10] Chamberlain C C,Huda W,Hojnowski L S,Perkins A and Scaramuzzino A 2000 Radiation doses to patients undergoing scoliosis radiography *Br J Radiology* **73** 847
[11] Wieder S and Adams P L 1981 Improved routine chest radiography with a trough filter *AJR* **137** 695
[12] Watanabe P C A,Issa J P M,Pardini L C,Monteiro S A C and Catirse A B C E B, 2007 A Singular method to compare dental radiographic films used to study maxillofacial structures *Int. J. Morphol.* **25** 573
[13] Kitagawa H and Farman A 2004 Effect of beam energy and filtration on the signal-to-noise ratio of the Dexis intraoral X-ray detector *Dentomaxillofacial Radiology* **33** 21

Brain activation in response to randomized visual stimulation as obtained from conjunction and differential analysis: an fMRI study

N H Nasaruddin[1, 2], A N Yusoff[1] and S Kaur[1]

[1] Universiti Kebangsaan Malaysia, Jalan Raja Muda Abdul Aziz, Kuala Lumpur, 50300, Malaysia
[2] Universiti Metropolitan Asia, Cheras, Selangor, 43200, Malaysia

E-mail: hanimnasaruddin@gmail.com

Abstract. The objective of this multiple-subjects functional magnetic resonance imaging (fMRI) study was to identify the common brain areas that are activated when viewing black-and-white checkerboard pattern stimuli of various shapes, pattern and size and to investigate specific brain areas that are involved in processing static and moving visual stimuli. Sixteen participants viewed the moving (expanding ring, rotating wedge, flipping hour glass and bowtie and arc quadrant) and static (full checkerboard) stimuli during an fMRI scan. All stimuli have black-and-white checkerboard pattern. Statistical parametric mapping (SPM) was used in generating brain activation. Differential analyses were implemented to separately search for areas involved in processing static and moving stimuli. In general, the stimuli of various shapes, pattern and size activated multiple brain areas mostly in the left hemisphere. The activation in the right middle temporal gyrus (MTG) was found to be significantly higher in processing moving visual stimuli as compared to static stimulus. In contrast, the activation in the left calcarine sulcus and left lingual gyrus were significantly higher for static stimulus as compared to moving stimuli. Visual stimulation of various shapes, pattern and size used in this study indicated left lateralization of activation. The involvement of the right MTG in processing moving visual information was evident from differential analysis, while the left calcarine sulcus and left lingual gyrus are the areas that are involved in the processing of static visual stimulus.

1. Introduction

Since the discovery of blood oxygenation level dependent (BOLD) effects [1], fMRI has become a standard experimental technique in accessing brain activity, among others, in response to visual stimuli [2]. This technique allows there presentation of visual field in human occipital cortex to be studied non-invasively [3]. The fMRI technique is also suitable for studying the properties of the visual area of subjects suffering from ocular abnormalities [4].

Human visual system has a limited processing capacity. Pervious information about visual image processing in human brain was obtained mainly from psychophysical studies [5]. Nowadays, fMRI technique has proven to be very useful in studying human brain function related to visual processing [6]. Visual processing system has to compete for cortical representation when multiple stimuli are simultaneously presented [7]. As a result, stimuli with a variety of features color or motion produces brain activation that are more prominent than stimuli that did not have any specific features [8].

In a study using a moving action visual stimulus [9], it was found that activation occurs in premotor cortex, superior parietal lobe and superior temporal sulcus. This was discussed as due to the fact that human brain can anticipate the movement as a result of motor stimulation. In other words, human brain may have the ability to simulate the performance of action being observed [10].

To date, a vast number of study on how human brain responded to visual stimuli have been conducted. Among others are facial recognition [6], stimulus in motion [9] and attention and memory [11]. Accompanying these studies are those conducted on patients with ocular abnormalities such as glaucoma [12] and chronic bilateral vestibular failure [13].

Ocular abnormalities have apparent effects on brain activation. However, neuroimaging data about the ocular characteristics of healthy participants are still lacking to be compared with data for patients with ocular pathology. The objective of this multiple participants' functional magnetic resonance imaging (fMRI) studies was to investigate the brain activation characteristics of healthy participants when they view stimuli of various shapes, pattern and size. Specifically, this study will determine the common activated areas, the areas that are more active in processing moving stimuli than static stimuli and the area that are more active in processing static stimuli than moving stimuli. Data obtained from healthy participants will serve as baseline for future studies on patients with ocular abnormalities.

2. Method

2.1 Participants

Sixteen healthy adult participants (9 malesand7females) participated in this study. This study has been approved by the Institutional Ethics Committee of the Universiti Kebangsaan Malaysia (Ref No: NN-073-2011). All participants were above 40years old (mean =48.7years, standard deviation=6.4years) with normal vision and no history of ocular disorder. All participants gave their informed written consent. They were required to undergo a comprehensive vision test conducted by qualified optometrist at Ophthalmology Clinic in Selayang Hospital. Visibility of the participants should be better than 6/12 without full refractive correction.

2.2 fMRI scan

The fMRIscans were performed in the Department of Radiology, Universiti Kebangsaan Malaysia Medical Centre (UKMMC) using a 3-T Siemens Magnetom Verio scanner. A standard BOLD imaging protocol with gradient echo planar imaging (EPI) pulse sequence was used with imaging parameters as follows: acquisition time (TA)=3000ms, echo time (TE)=50ms,field of view (FOV) =192×192mm, flip angle (α) =90°, matrix size=3×3×3and slice thickness=3mm.

2.3 Stimuli

Visual stimuli generated by a computer were projected onto the translucent screen via a liquid crystal display (LCD) projector. The non-metallic screen was fixed inside the magnet room while both the projector and the computer were placed on the control panel. The Participants viewed the stimuli on the screen via a mirror that was attached to the head coil. Five types of stimuli were used in this study; i) growing ring (A), ii) rotating wedges (B), iii) hourglass/bowtie (CD), iv) quadrant arc (EF) and iv) checkerboard pattern (G), see Figure 1 and [14]. All stimuli have black-and-white checkerboard pattern. Each stimulus has an element of movement except stimulus G which is static. Stimulus C& D and E & F flipped when displayed to the participant.

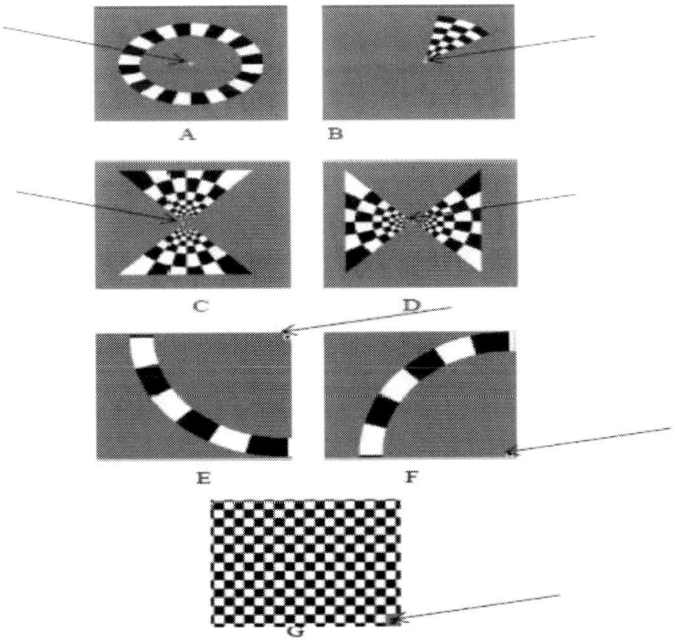

Figure 1: Visual stimuli used in this study. The participants were required to focus their sight onto a dot (indicated by the arrows) during the fMRI scan

2.4 fMRI paradigm and image processing

This study used a block-fMRI paradigm [15]. For the active state (stimulus shown), a number of 20 scan volumes were collected while for the rest state (participant viewing a white screen) a number of 10 scan volumes were collected. There were 80 scan volumes for each stimulus type and 160 scan volumes for rest state. The sum of all scan volumes is 480. All the A, B, CD, EF and G stimuli were projected in rotation during the active state. All participants were required to respond by squeezing the rubber bulb as they view the white screen. This is to ensure that the participants remain conscious and awake throughout the scans.

2.5 Post processing

The fMRI data were analyzed using MATLAB 7.4 – version R2010a (Mathworks Inc. MA, USA) and Statistical Parametric Mapping (SPM12b) programming software. Functional images in each measurement were realigned using the 6-parameter affine transformation in translational (x, y and z) and rotational (pitch, roll and yaw) directions. The normalization procedure used a 12-parameter affine transformation. The images were then smoothed using 8 an 8-mm full-width-at-half-maximum (FWHM) Gaussian kernel. Low-frequency responses caused by aliased biorhythms, cardiac effects and other oscillatory signal variations were removed using a high-passed filter. Conjunction analysis [16] was used to determine commonality of activation due to all stimulus types i.e. $A \wedge B \wedge CD \wedge EF \wedge G$. The brain activation was thresholded at $\alpha = 0.001$. The areas in the brain that were more activated for moving stimulus as compared to static stimulus, and vice versa, were determined using the combination of differential and conjunction analysis i.e. $(A > G) \wedge (B > G) \wedge (CD > G) \wedge (EF > G)$. The activation was also threshold at $\alpha = 0.001$. The activation areas obtained from the analyses above were confirmed using Anatomy Toolbox [17].

3. Results and Discussion

Figure 2 shows the activated areas obtained from conjunction analysis mentioned above. The activated areas are left precentral gyrus (*l*PCG), left post centralgyrus (*l*PostCG), left superior temporal gyrus (*l*STG), left occipital gyrus (*l*OG) and middle cingulate cortex (MCC).

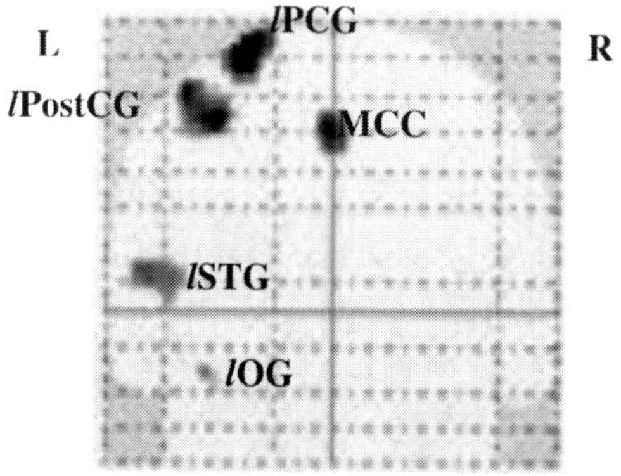

Figure 2: Maximum intensity projection in coronal orientation obtained from conjunction analysis, A Λ B Λ CD Λ EF Λ G, at α= 0.001. L: left, R: right

The results shown in Figure 2 indicated three important information when the participants viewed the visual stimuli of different shapes and size; i) the existence of common activated areas, ii) left lateralization of activation and iii) the existence of common activated areas outside visual cortex. The stimuli used in this study would be able to characterize visual world, visual field and visual space [14]. It is suggested that the variability of the function of stimuli and their unique ability to evoke responses in multiple brain areas have resulted in the activation in the visual cortex as well as outside the occipital cortex such as PCG, postCG, STG and MCC. The results also supported the functional integration concept which states that a particular brain area might have more than one function, for example, the PCG [18], postCG and STG [19] which have been found to be involved in working memory. The activation in MCC have not been reported to be specific for visual processing but it was found to be activated during learning processes [20], formation and processing of emotion [21] and working memory [22] which could occur during the task performance.

The involvement of right MTG in the processing of moving stimuli is evident in Figure 3(a) and is supported by previous study [9] which reported that MTG was significantly activated when the participants viewed capoeira movements and ballet dancing. MTG, also known as BA21, is located in between the STG and inferior temporal gyrus (ITG). The exact function of this area is inconclusive but is commonly related to distance estimation, face recognition and evaluation of meaning of words. It has also been related to motion perception and contrasting motion signals [23]. It has also been thought to play a major role in the perception of motion, the integration of local motion signals into global percept, and the guidance of eye movements [24].The brain areas that are more active towards static stimuli as compared to moving stimuli are left calcarine sulcus and left lingual gyrus (Figure 3(b)).The calcarine sulcus is where the primary visual cortex (V1) is concentrated. V1 has a very well-defined map of the spatial information. Participants in this study viewed complex visual stimuli thus activating fusiform, lingual and occipital gyrus significantly. Lingual gyrus has been known to play an important role in vision, recognition and identification [25].

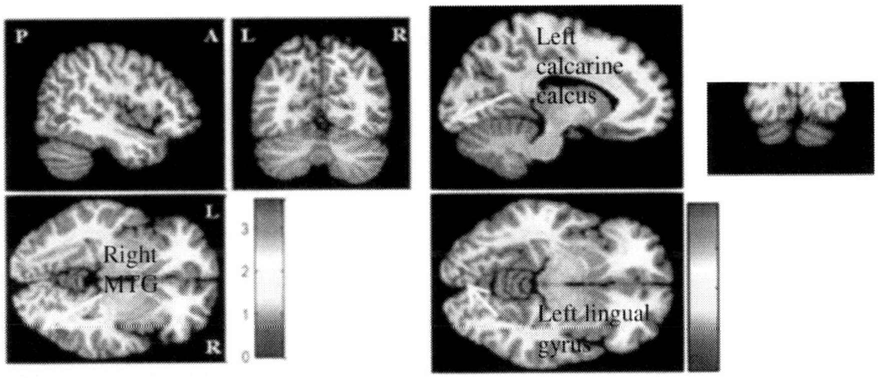

Figure 3: Blobs in red indicate significant activation at $\alpha = 0.001$ for brain area that are a) more active towards moving stimuli as compared to static stimuli and b) vice versa. P: Posterior, A: Anterior, L: Left, R: Right

4. Conclusion

Visual stimulation of various shapes, pattern and size used in this study indicated left lateralization of activation. The involvement of the right MTG in processing moving visual information was evident from differential analysis, while the left calcarine sulcus and left lingual gyrus are the areas that are involved in the processing of static visual stimulus.

Acknowledgement

This work was supported by UKM-GUP-2011-125 research grant, the Ministry of Education Malaysia and BKBP-FSK-K010146 research grant of Universiti Kebangsaan Malaysia.

References

[1] Ogawa S, Lee TM, Nayak AS,Glynn P 1990 *Magn Reson Med* **14** 68-78
[2] Grill-Spector K, Malach R 2004 *Annu Rev Neurosci* **27** 649-77
[3] Dumoulin SO, Wandell BA 2008 *Neuroimage* **39** 647-660
[4] Warnking J, Dojat M, Guérin-Dugué A, Delon-Martin C, Olympieff S, Richard N, Chéhikian A,Segebarth C 2002 *Neuroimage* **17** 1665-1683
[5] Puce A, Allison T, Asgari M, Gore JC,McCarthy G 1996 *The Journal of Neuroscience* **16** 5205-5215 [6] Gratton C, Sreenivasan KK, Silver MA, D'Esposito M 2013 *The Journal of Neuroscience* **33** 6979-6989
[7] Serences JT, Yantis S. Selective visual 2006 *Trends Cogn Sci* **10** 38-45
[8] Melcher D, Papathomas TV,Vidnyanszky Z 2005 *Neuron* **46** 723-9
[9] Calvo-Merino B, Glaser DE, Grezes J, Passingham RE,Haggard P 2005 *Cereb Cortex* **15** 1243-9
[10] Jeannerod M 1994 *Can J Physiol Pharmacol* **72** 535-41
[11] Sorstedt E, Persson A, Noren B, Bjornlert U, Malcherek P, Axelsson M, Johansson J,Smedby O 2005 *Acta Radiol* **46** 671-8
[12] Duncan RO, Sample PA, Weinreb RN, Bowd C,Zangwill LM 2007 *Progress in retinal and eye research* **26** 38
[13] Dieterich M, Bauermann T, Best C, Stoeter P,Schlindwein P 2007 *Brain* **130** 2108-16
[14] Duncan RO, Sample PA, Weinreb RN, Bowd C,Zangwill LM 2007 *Prog Retin Eye Res* **26** 38-56
[15] Yusoff AN, Mohamad M, Hamid AIA, Wan Abdullah WAK, Hashim MH,Zulkifli NZ 2010 *Malaysian Journal of Medicine and Health Sciences* **6** 71-92
[16] Friston KJ, Holmes AP, Price CJ, Buchel C,Worsley KJ 1999 *Neuroimage* **10** 385-96
[17] Eickhoff SB, Stephan KE, Mohlberg H, Grefkes C, Fink GR, Amunts K,Zilles K 2005 *Neuroimage* **25**1325-35
[18] Yusoff AN 2013 **42** 765-771

[19] Manan HA, Franz EA, Yusoff AN,Mukari SZMS 2013 *Neurology Psychiatry and Brain Research* **19** 180-191
[20] Hayden BY, Platt ML 2010 *J Neurosci* **30** 3339-46
[21] Hadland KA, Rushworth MF, Gaffan D,Passingham RE 2003 *Neuropsychologia* **41** 919-31
[22] Rushworth MF, Hadland KA, Gaffan D,Passingham RE 2003 *J Cogn Neurosci* **15** 338-53
[23] Takemura H, Ashida H, Amano K, Kitaoka A,Murakami I 2012 *J Neurosci* **32** 14344-54
[24] Born RT, Bradley DC 2005 *Annu Rev Neurosci* **28** 157-89
[25] Machielsen WC, Rombouts SA, Barkhof F, Scheltens P,Witter MP 2000 *Hum Brain Mapp* **9** 156-64

Radio-frequency coil selection for MR imaging of the carotid vessel wall

S Mat Isa[1], I L Shuaib[1] and S Bauk[2]

[1] Oncology and Radiological Science Cluster, Advanced Medical and Dental Institute, Universiti Sain Malaysia, Bertam, 13200 Penang, Malaysia
[2] Physics Section, School of Distance Educations, USM, Penang, Malaysia

E-mail: suzana@amdi.usm.edu.my

Abstract. This aim of this study was to identify the radiofrequency coil that will produce optimum image quality for scanning the carotid vessel wall using magnetic resonance imaging. A comparative cross-sectional study was conducted using 10 volunteers. Each volunteer was scanned three times using a 1.5T Signa HDxt machine equipped with one of three different coils: a neurovascular array (NV) coil, an 8-channel CTL spine array coil, and a 3-inch surface coil. A qualitative image quality rating was assigned to each image. The images were also evaluated by measuring the signal to noise ratio (SNR) using Osirix 4.2.3 software. The noise was estimated from the mean intensities of the region of interest in the background of the images and the signal was measured in the muscle adjacent to the vessel wall. The SNRs of the three coils were compared using one-way ANOVA, with 104 images used for the data analysis. The mean image quality scores for the NV head coil, CTL coil, and 3-inch coil were 3.4, 3.33, and 1.67, respectively. In addition, the SNRs differed significantly ($p < 0.05$). The mean SNR for the 3-inch coil was significantly higher (56.21 ± 25.06) than those for the NV head coil (27.34 ± 15.47) and CTL coil (21.77 ± 13.14). The Bonferroni post-hoc test revealed that there was no significant difference between the NV head coil and the CTL coil ($p = 0.21$). The optimum SNR value was 20-27. These results indicate that the NV head coil and CTL coil can be used to evaluate the carotid arterial wall with optimum image quality and higher resolution. These coil can deliver fast and robust data to image the carotid vessel wall in vivo.

1. Introduction

High spatial resolution magnetic resonance imaging (MRI) is a promising technique for assessing carotid atherosclerosis because it allows direct visualization of the diseased vessel wall and can be used to characterize the morphology of individual atherosclerotic carotid plaques [1]. MRI has been used to the study the human carotid artery [2], to diagnose atherosclerosis [3, 4], and to measure atherosclerosis [5, 6, 7]. Different types of radiofrequency (RF) coils can be used in MRI machines. In past studies designed to image arteriosclerosis in the carotid artery, researchers have used phased array coils [3, 4], custom-built phased array coils [5, 8], phased array surface coils [6, 7, 9, 10], and a 3-inch surface coil [11].

The MRI machine and the type of coil used can affect image quality. For example, Anumula et al. 2005 [12] compared the signal to noise ratio (SNR) and image quality obtained using dual-element phased array coils at 1.5 T and 3 T MRI machine. The mean SNRs for in vivo studies obtained at 3 T (17.6 ± 1.8 and 14.8 ± 2.7 for the left and right coils, respectively) differed from those at 1.5 T (8.6 ± 1.1 and 7.0 ± 1.0). The aim of this study was to identify the type of coil that will produce optimum image quality for scanning the carotid vessel wall. High quality images are crucial to diagnose arteriosclerosis and measuring the diameter of the artery.

2. Materials and Methods

2.1. Study population

Ten healthy volunteers (5 males and 5 females) ranging in age from 22 to 55 years old were recruited from the Advanced Medical and Dental Institute (AMDI) staff. The volunteers were asymptomatic with no history of stroke, hyperlipidemia, hypertension, or diabetes. Informed consent was obtained from all volunteers. The consent forms and the imaging protocols were approved by the Research Ethics Committee (Human), Universiti Sains Malaysia.

2.2. MRI examination

The volunteers were scanned using a Signa HDxt 1.5 T (GE Healthcare, Wisconsin, USA), and three different coils were tested: a neurovascular array (NV) coil (Invivo Corporation, Pawaukee, Wisconsin), a 3-inch surface coil (GE Medical Systems, Milwaukee, Wisconsin, USA), and an 8-channel CTL spine array coil (USA Instruments, Aurora, OH, USA). A 3D time-of-flight (TOF) sequence was used to identify the carotid bifurcation, and images were taken 1 cm above it. 2D fast spin echo (FSE) images with two contrast weightings (T1w and T2w) were obtained. The total length of the coverage was 80 mm. Table 1 summarizes the imaging parameters used in this study. Flow compensation was used to suppress flow artifacts. A double inversion recovery (IR) technique with cardiac gating was used to generate double IR proton density (PD) plaque images.

Table 1: Imaging parameters.

[a]Image parameters	TOF	T1w	T2w	Double IR PD Plaque
TR (ms)	16	400	3000	1500
TE (ms)	4	10	100	55
TI (ms)	–	–	–	Gated
Thickness (mm)	2	4	4	4
Matrix	256 x 160	256 x 256	256 x 256	256 x 256
NEX	1	4	4	4
FOV (cm)	16 x 16	16 x 16	16 x 16	16 x 16

[a]TR = Repetition time, TE = Echo time, TI = Inversion time, NEX = number of excitations, FOV = field of view.

2.3. Image evaluation

An image quality rating was assigned to each image following Zhang et al. [8]. Using their scale, the quality of each image from the three contrast weightings for every patient was evaluated by one reviewer. The images were excluded from the analysis due to motion artifacts if one or more of the two images (T1w and T2w) had a rating lower than three. The images also were evaluated quantitatively by measuring the SNR using Osirix 4.2.3 software (Figure 1). SNR is defined as the ratio of signal to image noise. The signal (S) was computed by drawing a region of interest (ROI) in the muscle adjacent to the vessel wall. This location was chosen because the vessel wall was only 1.5 mm thick, thus the assessment of the vessel wall SNR was prone to errors [12]. The noise standard deviation (SD) measurement was computed from the background region of the image in an area free of any visible artifacts. The SNRs calculated for the three coils were compared using one-way ANOVA.

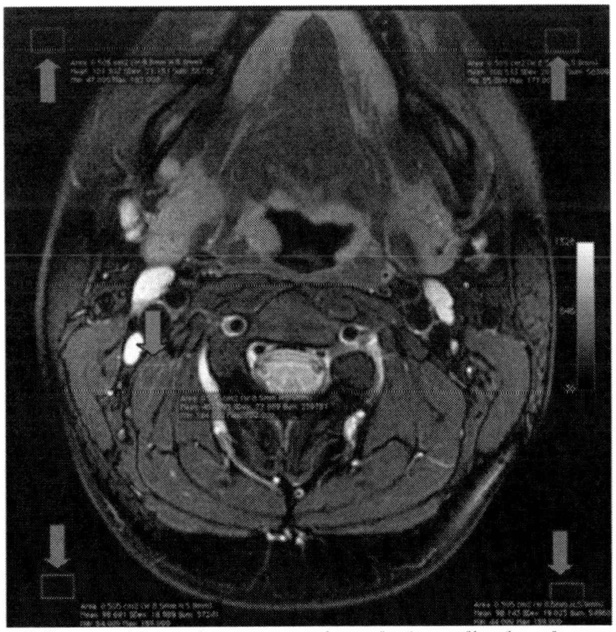

Figure 1: Axial image of the carotid arteries at approximately 1 cm distal to the carotid bifurcation in one subject. The signal was measured in the ROI, which was located in adjacent muscles. The noise level was measured by calculating the mean of the signal from four ROIs in the image background.

3. Results

Overall, 180 images were available for analysis. However, after qualitative analysis, 76 images with an image quality score less than 3 were excluded from the data review, leaving 104 images for SNR measurement. However, out of 104 images, the 3-inch coil was able to acquire only 16 usable images, whereas NV head coil and CTL coil were able to produce 44 usable images each. Figure 2 shows an example of good (A, B & C) and bad images with motion artifacts (D). Mean image quality scores for the NV head coil, CTL coil, and 3-inch coil were 3.4, 3.33, and 1.67 respectively. The mean SNR for the 3-inch coil (56.21 ± 25.06) was significantly higher ($p < 0.05$) than those for the NV head coil (27.34 ± 15.47) and CTL coil (21.77 ± 13.14). The Bonferroni post-hoc test revealed that there was no significant difference between the NV head coil and CTL coil ($p = 0.21$).

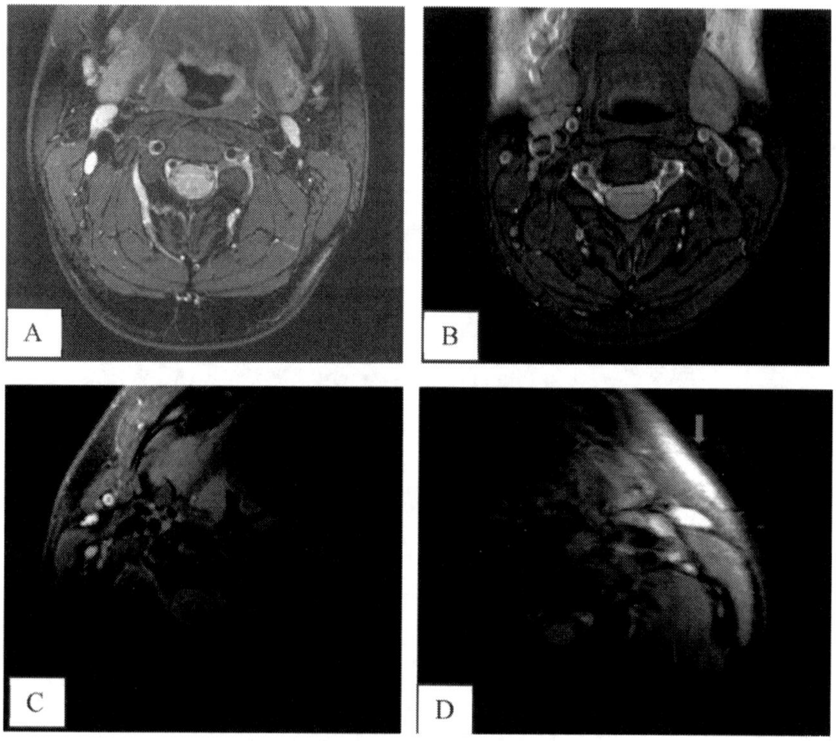

Figure 2: MRI images captured using the three different coils: A) NV head coil, B) CTL coil, and C) 3-inch coil. Motion artifacts (arrow) shown on 3-inch coil image (D).

4. Discussions

Quality of MRI images of the carotid vessel is crucial for diagnosing arteriosclerosis and taking quantitative measurements of the carotid artery wall. The first step to achieve high quality images is to identify the most suitable coil for use in the MRI machine. Several factors can affect image quality, such as patient motion, including swallowing and respiration; distance of the carotid artery from the coil, especially in obese patients; and incomplete flow suppression [8]. The factors that affect the SNR include magnetic field strength of the system, proton density of the area under examination, voxel volume, repetition time (TR), echo time (TE), flip angle, and number of excitations (NEX) [13]. In this study, most of these factors were standardized so that we could focus on identifying the type of coil that provided the best images.

Figure 3 shows the SNR and image quality score in these three coils. Note that, although the SNR in 3-inch coil is significantly higher ($p < 0.05$) than NV head and CTL coil, image quality score in 3-inch coil is low compared to the other coils. This due to the SNR equation, as it written as SNR = S/SD noise. Thus, when the SD of the noise is low, the SNR will be high. In this study, the mean SD of noise for the 3-inch coil was 9.79 ± 16.56, whereas it was 52.26 ± 23.44 for the NV head coil and 46.30 ± 22.14 for the CTL coil. At first glance the low noise value for the 3-inch coil seems advantageous, but it was outweighed by poor image quality. The mean image quality score for the 3-inch coil was 1.67 compared to 3.4 for the NV head coil and 3.33 for the CTL coil. The poor quality was due to artifacts caused by subjects' movement (figure 2), as they were not comfortable with the positioning required by 3-inch coil.

The subjects had to tilt their head when this coil was put in place, and they had to remain still until the imaging scan was completed, compared to normal supine position when the NV head coil and CTL coil were used, which made them comfortable during the scanning process. In fact, only 16 out of 60 images acquired using 3-inch coil is usable compared to 44 out of 60 images acquired using NV head and CTL coil are usable. A high SNR is commonly considered to be one of the major advantages of MRI. However, in this study, the optimum SNR was founded to be 20-27. This agreeable with Donald et al. (2007) that a SNR higher than 20 offers little in terms of improved image quality[14]. Thus, we can conclude that NV head with SNR of 27.34 ± 15.47 and CTL coils with SNR of 21.77 ± 13.14 can deliver optimum image quality to evaluate the carotid arterial wall for diagnosis of arteriosclerosis.

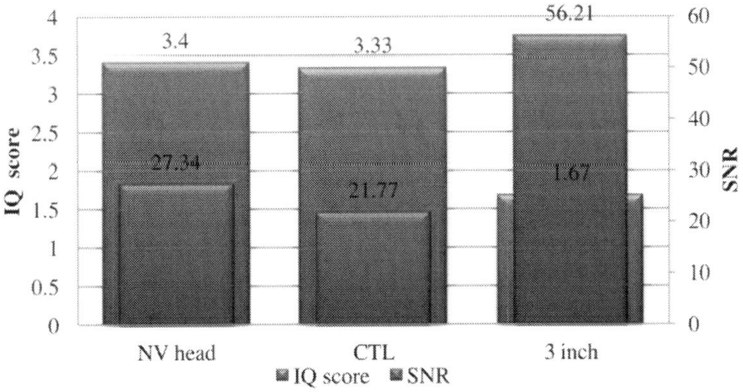

Figure 3: SNR and image quality score.

Furthermore, there are limitations of the 3-inch coil as it is not reproducible as only 1 out of 4 images will be usable, which is not practical in clinical setting. The acquisition of the coil is unilateral compared that bilateral in NV head and CTL coil. As a surface coil, this coil also only covers small area compared to NV head and CTL coil which are phased array coils. Due to carotid artery superficial location and length, it is also well suited for imaging using a phased array coil [15]. In addition, 3-inch coil also give less comfort to the patients. However, these results are preliminary and cannot be applied in a general case. A multi-centre study of adequate sample size would be required to implement these parameters in clinical settings.

5. Conclusion

In summary, the NV head coil and CTL coils can provide optimum image quality to evaluate the carotid arterial wall for diagnosis of arteriosclerosis. This coil can deliver fast, reproducible, and robust data and available in the most clinical setting.

Acknowledgement
This research was supported by a Universiti Sains Malaysia Short Term Grant (304/CIPPT/639041).

References

[1] Yuan C, Mitsumori LM, Beach KW, Maravilla KR 2001 *Radiology* **221** 285–99.

Journal of Physics: Conference Series **546** (2014) 012004 doi:10.1088/1742-6596/546/1/012004

[2] Hadley JR, Roberts JA, Goodrich KC, Buswell HR, Parker DL 2005 *Magn Reson Imaging* **23** 629–39.
[3] Corti R, Fayad ZA, Fuster V, Worthley SG, Helft G, Chesebro J, et al 2001 *Circulation* **104** 249
[4] Corti R, Fuster V, Fayad ZA, Worthley SG, Helft G, Smith D, et al. 2002. *Circulation* **106** 2884.
[5] Yuan C, Beach KW, Smith LH, Hatsukami TS 1998 *Circulation* **98** 2666–71.
[6] Kerwin W, Hooker A, Spilker M, Vicini P, Ferguson M, Hatsukami T, et al. 2003 *Circulation* **107** 851.
[7] Cai J, Hatsukami TS, Ferguson MS, Kerwin WS, Saam T, Chu B, et al. 2005 *Circulation* **112** 3437–44.
[8] Zhang S, Hatsukami TS, Polissar NL, Han C, Yuan C. 2001 *Magn Reson Imaging* **19** 795–802.
[9] Greenman RL, Wang X, Ngo L, Marquis RP, Farrar N. 2008 *Magn Reson Imaging* **26**:246–53.
[10] Saam T, Yuan C, Chu B, Takaya N, Underhill H, Cai J, et al. 2007 *Atherosclerosis* **194**.
[11] Wasserman BA, Smith WI, Trout HH, Cannon RO, Balaban RS, Arai AE. 2002 *Radiology* **223** 566–73.
[12] Anumula S, Song HK, Wright AC, Wehrli FW. 2005 *Acad Radiol* **12** 1521–6.
[13] Westbrook C. RCK& TJ 2005 *MRI in Practice 3rd Edition.* (United Kingdom: Blackwell Publishing)
[14] Donald W. M., Elizabeth A. M., Martin J. G. MRP 2007 *MRI from Picture to Proton* (New York, NY: Cambridge University Press)
[15] Quick HH, Debatin JF, Ladd ME. 2002 *Eur Radiol* **12** 889–900.

Journal of Physics: Conference Series **546** (2014) 012005 doi:10.1088/1742-6596/546/1/012005

Hand grips strength effect on motor function in human brain using fMRI: a pilot study

S S Ismail[1], M Mohamad[1], S O Syazarina [2] and W Y Nafisah [3]

[1] Diagnostic Imaging & Radiotherapy Program, School of Diagnostic and Applied Health Sciences, Faculty of Health Sciences, Universiti Kebangsaan Malaysia, 50300 Kuala Lumpur, Malaysia
[2] Department of Radiology, Universiti Kebangsaan Malaysia Medical Center, 56000 Cheras, Malaysia
[3] Department of Neurology, Universiti Kebangsaan Malaysia Medical Center, 56000 Cheras, Malaysia

E-mail: ctsuhana87@gmail.com

Abstract: Several methods of motor tasks for fMRI scanning have been evolving from simple to more complex tasks. Motor tasks on upper extremity were applied in order to excite the increscent of motor activation on contralesional and ipsilateral hemispheres in brain. The main objective of this study is to study the different conditions for motor tasks on upper extremity that affected the brain activation. Ten healthy right handed with normal vision (3 male and 7 female, age range=20-30 years, mean=24.6 years, SD=2.21) participated in this study. Prior to the scanning, participants were trained on hand grip tasks using rubber ball and pressure gauge tool outside the scanner. During fMRI session, a block design with 30-s task blocks and alternating 30-s rest periods was employed while participants viewed a computer screen via a back projection-mirror system and instructed to follow the instruction by gripping their hand with normal and strong grips using a rubber ball. Statistical Parametric mapping (SPM8) software was used to determine the brain activation. Both tasks activated the primary motor (M1), supplementary motor area (SMA), dorsal and ventral of premotor cortex area (PMA) in left hemisphere while in right hemisphere the area of primary motor (M1) somatosensory was activated. However, the comparison between both tasks revealed that the strong hand grip showed the higher activation at M1, PMA and SMA on left hemisphere and also the area of SMA on right hemisphere. Both conditions of motor tasks could provide insights the functional organization on human brain.

1. Introduction

Since the number of various studies relates to usage of functional Magnetic Resonance Imaging (fMRI) has been increasing owing to non-ionization modality, it implies an importance in accommodating the enhancement of motor action for clinical research such as reorganization of motor recovery and plasticity for stroke patients [1]. Several methods of motor tasks for fMRI scanning have been evolving from simple to more complex tasks such as finger tapping [1-4], hand grip [5-8] and wrist extension [9-10]. Motor tasks on upper extremity were applied in order to excite the increscent of motor activation in the brain [5-9]. These motors action usually revealed the activated areas and connectivity not only in the contralateral hemisphere but also in the ipsilateral hemisphere [11-12].

The common areas which activated in the brain for motor task are the primary motor (M1) [1], supplementary motor (SMA) [13-15], premotor cortex (PMA) [16-18], and somatosensory areas.

The importance of the study is to highlight the tasks which suitable for stroke patient and the information about functional specialization of motor activation might appropriate task to be use in the rehabilitation of strokes patients and pave the way for future studies in fMRI. The aim of this study was to identify the brain areas (activation) of motor and brain activation (% percentage of signal changes) of functional areas between normal and strong hand grips. We hypothesized that different strength of hand grips would give different activation areas of motor in brain.

2. Material and methods

2.1. Subject

This quasi -experimental study was done using the fMRI of 3 Tesla (3T) at Department of Radiology, Universiti Kebangsaan Malaysia Medical Centre (UKMMC). This study was approved by the Universiti Kebangsaan Malaysia (UKM) Ethics Committee (IEC) (NN-051-2013). Ten healthy volunteers participated in this study (3 males and 7 females, age range=20-30 years, mean=24.6 years, SD=2.21) with no neurological or physiatrist diseases. All participants were right-handed tested by Edinburg handedness inventory [19]. They were given written consent and screening forms prior to scanning.

2.2. Experimental Paradigm

Participants were trained outside the scanner to perform the hand grip exercise using a rubber ball and pressure gauge tool. This is to measure their normal and strong grip before entering the MRI. This task is important for participant in order to make sure they know and familiar with the tasks given during the scanning. Overall, pressure for normal grip are between 3-5 psi and for strong grip are approximately between 5-10 psi.

The participants performed a grip task with the right hand in the presence of visual cues by applying normal and strong grip to a rubber ball. The block design paradigm (Figure 1) consists of 12 active (normal and strong grips) and 12 rest blocks, 30s each was used in the experiment.

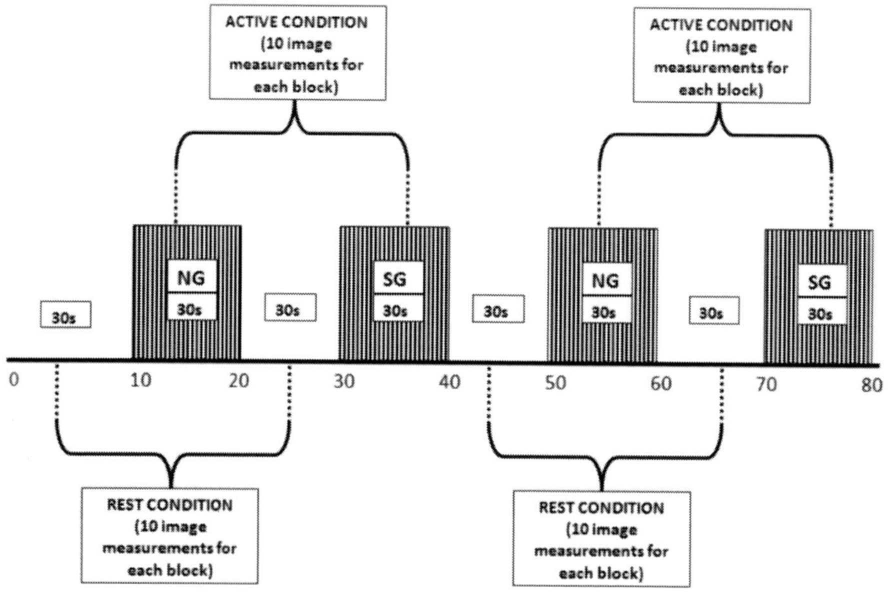

Figure 1. Block paradigm for hand grips motor task.

2.3. fMRI Data Analysis

All the functional (T2*-weighted) and structural (T1-weighted) images were analysed using Matlab R2012a (Mathworks Inc. Natrick, MA, USA) and Statistical Parametric Mapping (SPM8) software package shown in Figure 2.

Activated voxels were identified by pre-processing (realignment, normalization, and smoothing) by deriving the appropriated test statistic (T statistic) with corrected p-value =0.05 and post processing.

Percentage of signal changes (PSC) for regions of intereset (ROIs) with sphere of 5mm were analyzed using Marsbar toolbox [20].

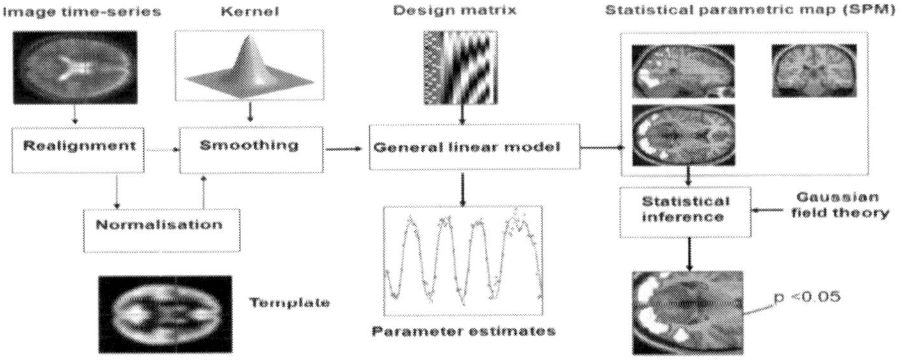

Figure 2. fMRI data analysis using SPM software.

3. Results

Brain map analysis (Figure 3 and Figure 4)of both tasks normal and strong grips showed activation in the primary motor (M1), supplementary motor area (SMA), dorsal and ventral of premotor cortex area (PMA) in left hemisphere and primary motor (M1) somatosensory in right hemisphere.

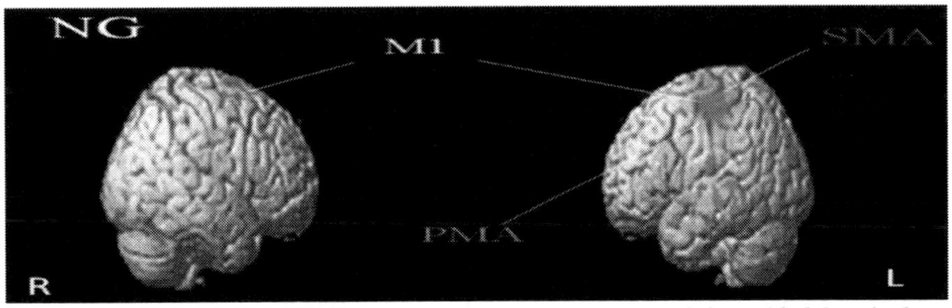

Figure 3. ROI of brain activation for normal grip (NG) on right and left hemispheres.

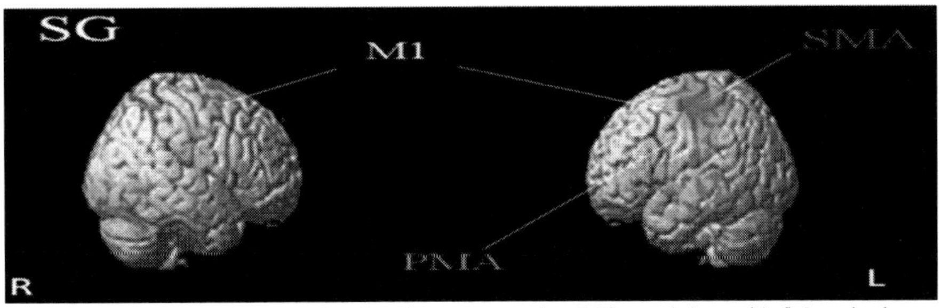

Figure 4. ROI of brain activation for strong grip (SG) on right and left hemispheres.

The percentages of signal changes (PSC) in Table 1 revealed that on left hemisphere the value of activated voxels were slightly significant between normal with strong grips. The areas of activation of M1 are higher followed by area of SMA and PMA on left hemisphere shown in Figure 4.

Table 1. The percentage signal changes (PSC) from group analysis in the ROI at M1, PMA and SMA for normal and strong grips on left hemisphere.

Left hemisphere	Normal Grip	Strong Grip
M1	2.053	2.460
SMA	1.876	2.269
PMA	1.887	1.728

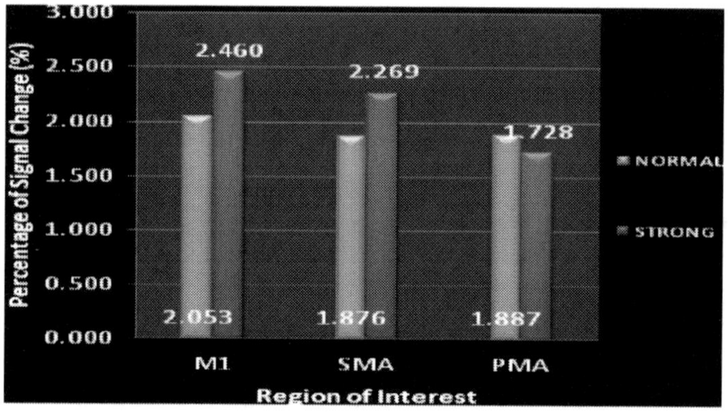

Figure 5. Percentage of signal change (%) of activated on motor areas for normal and strong grips on left hemisphere.

4. Discussion

From Figure 3 and Figure 4, it is interesting to see that both tasks normal and strong grips shown activation at the areas of M1, SMA and PMA in left hemisphere while in right hemisphere the area of M1 somatosensory was activated. This indicated the existence of ipsilaterality on right hemisphere accompanying the expected contralaterality on left hemisphere while doing the strong and normal grips task.

However, the comparison between both tasks revealed higher number of activated voxels and higher activation intensity in M1 and SMA for strong grip compared to the normal grips.

The strength of both tasks indicated that normal participants were able to do it at the range of 4 -5 psi for normal hand grips while for strong grips were between 7-10 psi. Review from all participants stated that these simple task are easily and comfortable to be applied during the scan for about 30 sec per block. Hence, simple tasks like these may provide suitable task for stroke patient.

From Table 1 and Figure 5, PSC were higher observed in the left M1 followed by SMA and PMA for strong grip task. This was due to higher demand of force doing the strong grip task compared to the normal grip task.

5. Conclusion

Strong grip task has higher percentage of signal changes compare to normal grip. This can verify that different strength of hand grips have affected motor areas of brain activation on left and right hemispheres. Both conditions of motor tasks could provide insights the functional organization on human brain and could contribute to medical decision making in neurological and rehabilitation treatment for stroke patient.

Acknowledgement
The authors would like to thank the staffs from the Department of Radiology, UKMMC. This work was supported by the UKM research GGPM-2012-107.

References
[1] Yusoff AN, Mohamad M, Hamid KA, Hamid AIA, Manan HA, Hashim MH 2010 *Activation characteristics of the primary motor (M1) and supplementary motor (SMA) areas during robust unilateral finger tapping task. Jurnal Sains Kesihatan Malaysia 8(2):43-9.*
[2] Yusoff AN, Hashim MH, Ayob MM, Kassim I 2008 *Pengaktifan otak akibat gerakan jari bagi subjek dominan tangan kanan dan kiri. J Sains Kes Mal 4(2):63-83.*
[3] Yusoff AN, Mohamad M, AI AH 2008 *The acquisition, analyses and interpretation of fMRI data: a study on functional specialisation in primary auditory cortices. Medicine & Health 3(2):300-17.*
[4] Yusoff AN, Mohamad M, Hamid KA 2011 *Pemerolehan, Analisis dan Interpretasi Data fMRI: Kehubungan Efektif dalam Korteks Auditori Primer Manusia. Sains Malaysiana 40(6):665-78.*
[5] Bestmann S, Swayne O, Blankenburg F, Ruff CC, Teo J, Weiskopf N, et al 2010 *The role of contralesional dorsal premotor cortex after stroke as studied with concurrent TMS-fMRI. The Journal of neuroscience 30(36):11926-37.*
[6] Marshall RS, Zarahn E, Alon L, Minzer B, Lazar RM, Krakauer JW 2009 *Early imaging correlates of subsequent motor recovery after stroke. Annals of neurology 65(5):596-602.*
[7] Nair DG, Hutchinson S, Fregni F, Alexander M, Pascual Leone A, Schlaug G 2007 *Imaging correlates of motor recovery from cerebral infarction and their physiological significance in well-recovered patients. Neuroimage 34(1):253-63.*
[8] Nowak DA, Grefkes C, Dafotakis M, Eickhoff S, Kust J, Karbe H, et al 2008 *Effects of low-frequency repetitive transcranial magnetic stimulation of the contralesional primary motor cortex on movement kinematics and neural activity in subcortical stroke. Archives of Neurology 65(6):741.*
[9] Dechaumont-Palacin S, Marque P, De Boissezon X, Castel-Lacanal E, Carel C, Berry I, et al 2008 *Neural correlates of proprioceptive integration in the contralesional hemisphere of very impaired patients shortly after a subcortical stroke: an fMRI study. Neurorehabilitation and Neural Repair 22(2):154-65.*
[10] Lindberg PG, Schmitz C, Engardt M, Forssberg H, Borg J 2007 *Use-dependent up-and down-regulation of sensorimotor brain circuits in stroke patients. Neurorehabilitation and Neural Repair.*
[11] Ward NS, Newton JM, Swayne OBC, Lee L, Thompson AJ, Greenwood RJ, et al 2006 *Motor system activation after subcortical stroke depends on corticospinal system integrity. Brain 129(3):809-19.*
[12] Wei W, Bai L, Wang J, Dai R, Tong RK-y, Zhang Y, et al 2013 *A Longitudinal Study of Hand Motor Recovery after Sub-Acute Stroke: A Study Combined fMRI with Diffusion Tensor Imaging. PloS one 8(5):e64154.*
[13] Ogawa S, Lee T, Kay A, Tank D 1990 *Brain magnetic resonance imaging with contrast dependent on blood oxygenation. Proceedings of the National Academy of Sciences 87(24):9868.*
[14] Lotze M, Markert J, Sauseng P, Hoppe J, Plewnia C, Gerloff C 2006 *The role of multiple contralesional motor areas for complex hand movements after internal capsular lesion. The Journal of neuroscience 26(22):6096-102.*
[15] Lotze M, Montoya P, Erb M, Hülsmann E, Flor H, Klose U, et al 1999 *Activation of cortical and cerebellar motor areas during executed and imagined hand movements: an fMRI study. Journal of Cognitive Neuroscience 11(5):491-501.*
[16] Cramer SC 2008 *Repairing the human brain after stroke: I. Mechanisms of spontaneous recovery. Annals of neurology 63(3):272-87.*

Journal of Physics: Conference Series **546** (2014) 012005 doi:10.1088/1742-6596/546/1/012005

[17] Ward N, Brown M, Thompson A, Frackowiak R 2003 *Neural correlates of outcome after stroke: a cross-sectional fMRI study. Brain 126(6):1430-48.*
[18] Ward NS 2004 *Functional reorganization of the cerebral motor system after stroke. Current opinion in neurology 17(6):725.*
[19] Oldfield RC 1971 *The assessment and analysis of handedness: the Edinburgh inventory. Neuropsychologia 9(1):97-113.*
[20] Brett M, Anton J-L, Valabregue R, Poline J-B 2002 *Region of interest analysis using the MarsBar toolbox for SPM 99. Neuroimage 16:S497.*

Journal of Physics: Conference Series **546** (2014) 012006 doi:10.1088/1742-6596/546/1/012006

fMRI of pain studies using laser-induced heat on skin with and without the loved one near the subject – a pilot study on 'love hurts'

T Sofina, W A Kamil and A H Ahmad

School of Medical Sciences, Universiti Sains Malaysia, 16150 Kubang Kerian, Kelantan, Malaysia

E-mail: sofina_tamam@yahoo.com

Abstract. The aims of this study are to image and investigate the areas of brain response to laser-induced heat pain, to analyse for any difference in the brain response when a subject is alone and when her loved one is present next to the MRI gantry. Pain stimuli was delivered using Th-YAG laser to four female subjects. Blood-Oxygenation-Level-Dependent (BOLD) fMRI experiment was performed using blocked design paradigm with five blocks of painful (P) stimuli and five blocks of non-painful (NP) stimuli arranged in pseudorandom order with an 18 seconds rest (R) between each stimulation phase. Brain images were obtained from 3T Philips Achieva MRI scanner using 32-channel SENSE head coil. A T1-weighted image (TR/TE/slice/FOV = 9ms/4ms/4mm slices/240×240mm) was obtained for verification of brain anatomical structures. An echo-planar-imaging sequence were used for the functional scans (TR/TE/slice/flip/FOV=2000ms/35ms/4mm slices/90°/220×220mm). fMRI data sets were analysed using SPM 8.0 involving preprocessing steps followed by t-contrast analysis for individuals and FFX analysis. In both with and without-loved-one conditions, neuronal responses were seen in the somatosensory gyrus, supramarginal gyrus, thalamus and insula regions, consistent with pain-related areas. FFX analysis showed that the presence of loved one produced more activation in the frontal and supramarginal gyrus during painful and non-painful stimulations compared to absence of a loved one. Brain response to pain is modulated by the presence of a loved one, causing more activation in the cognitive/emotional area i.e. 'love hutrs'.

1. Introduction

Pain is multidimensional and usually is very subjective with respect to individuals. Based on available literatures, pain is defined as the sensory and emotional experience of discomfort.[1] Pain may also be defined as the feeling or perception of irritating, miserable, or unbearable sensation arising from a part of the body.[2] Basically, all types of pain may result in the activations in common pain-related regions. More specifically, pain from experimental laser heat stimuli activates the areas of primary somatosensory,[3-5] secondary somatosensory,[3] insular cortex,[3-7] thalamus,[3,5] amygdala and anterior cingulate cortex.[3,4]

While there are many fMRI studies of pain, few have investigated the correlation between pain and the influence of love to pain response. Cheng et al, (2010) requested the subjects to imagine the loved one or an unfamiliar individual being in a painful situation. The fMRI data indicated that the perspective of a loved-one showed increased activity in the anterior cingulate cortex and insula.[8]

So far no known studies have investigated the effect of having a loved one physically nearby on brain activities while pain stimulation is being given. Since pain is related to emotion, this study is carried out to look for the influence of type of emotion i.e. love to the pain response. This study utilised the technique of fMRI to show the activation of the neuronal responses associated with pain. The main focus of the study is to analyse the difference in the hemodynamic response when each subject is alone and when her loved one is present next to the MRI gantry.

2. Method

All volunteers underwent standard procedure for fMRI scanning. Prior to scanning session, all the subjects, the loved ones and researcher have already taken the safety step to ensure they were eligible for MRI examination. Subjects and the loved ones have been cautioned on MR hazards and the precautions should they take. All individuals were asked to fill out the pre-screening form before entering the MR environment and room including the researcher. Ferrous materials on their body or in pocket were removed as well as make up and jewelleries. All of them were given earplugs and MR compatible electrostatic headphone to protect their hearing.

Each subject was provided with a response button and was instructed to focus and give full attention to response. A researcher stayed with the subject in the scanner room to hold the laser with its beam moving to make several spots on the dorsum of the subject's hand. The experiment was repeated twice but in different situation; (1) subject was alone in the MRI gantry (this situation is referred to as 'without-loved-one condition'); (2) subject with the loved one was present next to the gantry (this is referred to as 'with-loved-one condition'). Note that the 'without-loved-one condition' meant that the loved one was many miles away from the experiment setting. The loved one could be one of these candidates; mother, father, partner or best friend which was determined through the Experiences in Close Relationships - Relationship Structure questionnaire.[9]

2.1. Subject

Four right handed female participants (mean age 27.25; SD 0.5 years) involved in this study were MRI compliant and have signed informed consent form in compliance with the regulations of the Universiti Sains Malaysia. This research was approved by the Human Research Ethics Committee of USM in 2012. Ethical approval was obtained for all volunteers, their loved ones who accompany them in the scan room, and the researcher delivering the pain stimulation in the scan room. Participants were all healthy volunteers. Only females were selected to reduce confounding emotional effects and to minimise gender variability. This was based on the study that men and women are different in the type of pain experience and reactions to pain.[10] In addition, women are more sensitive to many sensory tests compared to men.[11]

2.2. Experimental Paradigm

An fMRI experiment was performed using a block design paradigm. Firstly, the paradigm was developed in E-Prime System so as to synchronise the timing with the MRI scanner. The experiment consisted of 2x5 stimulation blocks of 18 seconds. There were five blocks of painful stimuli (P) and five blocks of non-painful stimuli (NP) arranged in pseudorandom order. Each stimulation phase was followed by an 18 seconds rest (R) of no activity.

The subject needed to focus her attention on the stimuli without any specific discrimination task and to give response to each block of stimuli by pressing the response button. In order to avoid nociceptor fatigue or sensitisation, the laser beam spot was slightly moved after each stimulus.[12] On the second part of the experiment, the loved one that sat next to the MRI gantry was not allowed to touch or talk to the subject to avoid any sensory or audio stimulation which may initiate unnecessary responses.

2.3. Pain Stimuli

The pain stimuli was delivered using Th:Y AG laser from THEMIS laser device available in Hospital Universiti Sains Malaysia (HUSM). The laser with a wavelength of 2μm and a diameter of 5mm was targeted to skin via an MRI compatible applicator. In this research context, the painful stimuli means the laser beam with high energy enough to cause pain on the subject's skin while the non-painful stimuli means the laser beam with lower (control) energy that results in warm sensation but not yet reaching pain threshold.

2.4. MR Image Acquisition.

This study utilised the 3.0T Philips Achieva MRI scanner[13] available in the Department of Radiology, Hospital Universiti Sains Malaysia (HUSM) with a 32-channel SENSE head coil for pulse transmission and signal reception. For each subject, a T1-weighted, high resolution structural image (TR/TE/slice/FOV = 9ms/4ms/4mm slices/240×240mm)[13] was obtained for verification of brain anatomical regions. An echo-planar imaging (EPI) sequence with the following parameter (TR/TE/slice/flip angle/FOV = 2000ms/35ms/4mm slices/90°/220×220mm)[13] was used for functional scans. Both scans corresponded to a total scan time of approximately 11 minutes per subject.

2.5. Data Analysis

The data collected was sets of fMRI images in a time series. The researcher also observed the subject's response while she was in the MRI gantry. Image analysis was performed using the MATLAB 7.4 – R2007a (Mathworks Inc., MA, USA) and Statistical Parametric Mapping (SPM 8.0) (Wellcome Department of Imaging Neurosciences, Institute of Neurology, University College of London, UK) software packages. The pre-processing steps were applied to all fMRI data set to remove any data artefacts and to improve the signal-to-noise ratio (SNR) in the time series data acquired. The steps involved were realignment, normalisation and smoothing. The t-contrast comparison was carried out for individuals and group. Since there were only four subjects, the fixed-effect analysis (FFX) was used to see the difference between signals or within subject variability and not to represent a population. The coordinates of activations were matched with MNI coordinate system using wfupickatlas in SPM.

2. Result

The height threshold for individual analysis were in the range of $5.24<T<5.29$ ($p<0.05$(FWE)) for without-loved-one condition and $5.12<T<5.24$ ($p<0.05$(FWE)) for with-loved-one condition. While the height threshold for FFX analysis were $T=5.06$ ($p<0.05$(FWE)) for without-loved-one condition and $T=4.93$ ($p<0.05$(FWE)) for with-loved-one condition.

3.1. Observation

During scanning, subject's responses to stimuli were monitored. In the with-loved-one condition, three out of four subjects pressed the response button for almost all stimuli, including the non-painful stimuli.

3.2. Individual Analysis

Figure 1 showed activations in individual subjects in both conditions during painful (P) stimulation. There were clear activations found in the somatosensory gyrus, insula, supramarginal gyrus and thalamus.

When the loved one was present (Figure 1(b)), neuronal responses appeared in more regions in the brain but with lower intensities in all subjects (A, B, C and D). The activated area that was not found in without-loved-one condition (Figure 1(a)) was the cingulum.

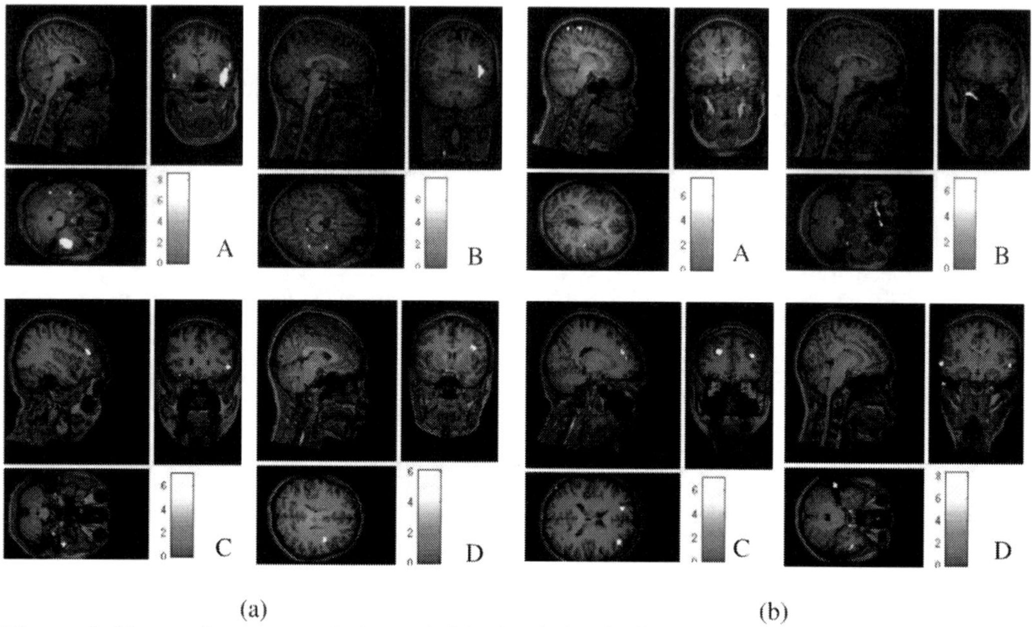

Figure 1. Neuronal responses during painful stimulation in four subjects A, B, C, and D in (a) the without-loved-one condition and (b) the with-loved-one condition.

Table 1. Summary of the areas of activation, coordinates and t-values in two different conditions for all stimulations.

	Without-Loved-One Condition			With-Loved-One Condition		
	Areas of activation	Coordinate (x y z) mm	t-values	Areas of activation	Coordinate (x y z) mm	t-values
P	Temporal-Sup	56 6 -2	6.72	Frontal-Mid	36 56 20	6.32
	Supramarginal	-60 -52 26	6.66	Temporal-Mid	68 -48 4	6.12
	Somatosensory	44 -28 34	6.41	Somatosensory	48 -36 62	6.08
	Thalamus	16 -12 16	6.19	Mid-Cingulate	0 16 40	5.92
	Supramarginal	66 -44 28	5.88	SMA	34 -12 66	5.67
	Somatosensory	68 -14 18	5.87	SMA	42 -18 64	5.59
	Insula	68 -36 26	5.79	SMA	-50 12 34	5.58
	Supramarginal	-38 16 -4	5.61	Insula	40 8 4	5.50
NP	Supramarginal	-60 -52 28	7.90	Temporal-Sup	64 2 4	6.94
	Frontal-Inf	48 22 10	7.37	Somatosensory	46 -36 62	6.16
	Parietal-Inf	58 -56 46	6.92	Frontal-Mid	36 56 20	6.00
	SMA	0 16 66	6.73	SMA	-48 4 42	5.73
	Somatosensory	48 -24 32	6.71	Frontal-Sup	22 -12 72	5.68
	Insula	-40 16 -4	6.32	Mid-Cingulate	0 16 40	5.66
R	Supramarginal	-60 -52 26	7.61	Frontal-Sup	-20 60 24	5.60
	Parietal-Inf	58 -56 46	6.57	Frontal-Inf	48 28 -18	5.53
	Frontal-Mid	26 60 22	6.54	Rectus	-8 58 -18	5.52
	Frontal-Inf	46 22 14	6.26	Temporal-Sup	56 14 -16	5.21

P = painful stimulation, NP = non-painful stimulation, R = rest
Temporal-Sup = superior temporal gyrus, Temporal-Mid = middle temporal gyrus, Frontal_Mid = middle frontal gyrus, Frontal-Inf = inferior frontal gyrus, Frontal-Sup = superior frontal gyrus, Parietal-Inf = inferior parietal gyrus, SMA = sensory motor area

3.3.2. Painful vs Rest (P-R). In the without-loved-one condition, the differences were found in SMA, precuneus and cuneus (Figure 2). These were the regions more activated during painful stimuli compared to during rest. In the with-loved-one condition, more differences were found in frontal and temporal areas, sensory motor area and mid-cingulate (mCC).

(a) (b)

Figure 2. The brain signal in the painful vs rest comparison for (a) without-loved-one condition and (b) with-loved-one condition.

3.3.3. Non-Painful vs Rest (NP-R). There was no difference found during non-painful stimuli compared to rest in the without-loved-one condition. However, many activations were found in the with-loved-one condition including in the SMA, supramarginal and mid-cingulate (mCC) as shown in Figure 3.

(a) (b)

Figure 3. The brain signal in the Non-Painful vs Rest Comparison for (a) Without-Loved-One Condition and (b) With-Loved-One Condition.

3.3.4. Painful vs Non-Painful (P-NP). In the without-loved-one condition, the differences were seen in somatosensory gyrus (SI) and cuneus as shown in Figure 4. (a). However in Figure 4. (b), there was no difference in the brain activation between painful and non-painful stimuli in the with-loved-one condition.

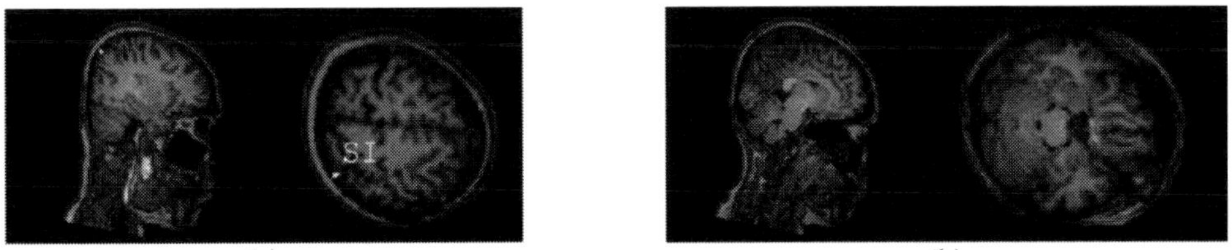

(a) (b)

Figure 4. The brain signal in the Painful vs Non-Painful Comparison for (a) Without-Loved-One Condition and (b) With-Loved-One Condition

3.3.5. Non-Painful vs Painful (NP-P). This comparison did not give significant result.

3.3.6. Direct Contrast Comparison. Table 2 shows the comparisons between the with-loved-one and without-loved-one conditions during all stimulations. During painful stimulation, presence of loved one produced more activities in frontal gyrus and right supramarginal gyrus compared to when the loved one was absent. The non-painful stimulation produced activities in right and left supramarginal, frontal and temporal regions. On the other hand, the contrast with vs without loved one did not produce any significant activations.

Table 2. The direct contrast comparison between painful and non-painful stimulations for both with and without-loved-one conditions.

Pain Stimulation	Contrast Comparisons	Coordinate of activations	Anatomical Area
P-R	(with vs without) loved one	40 54 22	Frontal-Mid
		34 58 14	Frontal-Mid
		60 -36 36	Right Supramarginal
	(without vs with) loved one	-	-
NP-R	(with vs without) loved one	60 -34 36	Right Supramarginal
		-62 -48 20	Temporal-Sup
		62 -48 28	Left Supramarginal
		-38 50 20	Frontal-Mid
	(without vs with) loved one	-	-

4. Discussion

The current study demonstrates that pain stimulation in all subjects activated pain-related areas similar to that observed in previous studies. These areas include the somatosensory gyrus, insula cortex and thalamus[3-8,12].

From the behavioural observation, three out of four subjects tended to show more pain when the loved one was with them. This reaction maybe compared to a study which found that patient who felt their spouses was solicitous reported more pain when their spouses watched than when the employee (stranger) did.[14] A similar reaction may occur even when the parents or the family members were around. The solicitousness of the loved one may lead to more pain behaviour.[15]

The obvious difference in FFX analyses between the conditions, with and without loved one, seems to suggest evidence of 'love hurts'. Non-painful stimuli should not lead to pain experienced by the subject; however, the activation pattern during non-painful stimuli was similar to that of during painful stimuli in with-loved-one condition. With loved-one present, a warm sensation can result in pain to some individuals. In the without-loved-one condition, on the other hand, the results showed that non-painful stimulation i.e. warm sensation gave no pain to subject, just like during rest. In the with-loved-one condition, the love hurts evidence was stronger when the comparison of painful and non-painful stimuli showed no difference. It meant that when the loved one stayed near the subject, warm or pain stimuli was similarly felt. This study only used four subjects. Results should become better when more participants are added.

By comparing contrast areas between the conditions, the with-loved-one condition indicated more activation. With the presence of a loved one, painful as well as non-painful stimulations activated the supramarginal gyrus and frontal gyrus. In contrast, in Cheng's work,[8] the presence of loved one was associated with stronger neuronal activation in cingulate cortex. Funtionally, frontal gyrus is associated with cognition and emotion as well as attention to pain while the supramarginal gyrus appears to play a central role in controlling our empathy towards other people.[16]

5. Conclusions

BOLD fMRI technique utilising the experimental paradigm for pain studies developed for this work has been able to map different neuronal activations in different parts of the brain when the subject is with the loved one and when she is not with the loved one near the gantry. Overall, this study indicates that the love hurts phenomena can be the reason for an individual to experience more pain when the loved one was around.

Acknowledgement
The authors would like to thank the Department of Radiology, Hospital Universiti Sains Malaysia (HUSM) for the permission to use the MRI scanner.

References

[1] American Medical Association, AMA 2003 In a Sarafino E D 2006 *Health Psychology: Biopsychosocial Interaction (5 th ed.)*. New Jersey John Wiley & Sons, Inc.

[2] Bear M F, Connors B W and Paradiso M A 2007 *Neuroscience Exploring The Brain (3rd ed.)*. Baltimore : Lippincott.

[3] Apkarian A V, Bushnell M C, Treede R D and Zubieta J K 2005 *Human Brain Mechanism of Pain Perception and Regulation in Health and Disease* European Journal of Pain pp 463-484.

[4] Bornhovd K, Buchel C, Quante M, Bromm B and Weiler C 2002 *Painful Stimuli Evoke Different Stimulus – Response Functionin the Amygdala, Prefrontal, Insula and Somatosensory Cortex : A Single-Trial fMRI Study*. Brain 125 pp 1326 – 1336.

[5] Bingel U, Quante M, Knab R, Bromm B, Weiller C and Buchel C 2003 *Single Trial fMRI Reveals Significant Contralateral Bias in Responses to Laser Pain Within Thalamus and Somatosensory Cortices*. Neuroimage 18 pp 740–8.

[6] Peyron R., Laurent B and Garcia-Larrea L 2000 *Functional Imaging of Brain responses to Pain. A Review and Meta-analysis*. Neurophysiology Clinical, 30 pp 263 – 288.

[7] Baumgärtner U, Iannetti G D, Zambreanu L, Stoeter P, Treede R D and Tracey I 2010 *Multiple Somatotopic Representations of Heat and Mechanical Pain in the Operculo-Innsular Cortex: High Resolution fMRI Study*. Journal of Neurophysiology 104 pp 2863 – 2872.

[8] Cheng Y, Chen C, Lin C P, Chou K H and Decety J 2010 *Love Hurts: An fMRI Study* Neuroimage 51 pp 923-929.

[9] Fraley R C, Heffernan M E, Vicary A M and Brumbaugh C C 2011 *The Experiences in Close Relationship – Relationship Structures Questionnaire: A Method for Assessing Attachment Orientations Across Relationships*. Psychological Assessment 23 pp 615-625.

[10] Sarafino E D 2006 *Health Psychology. Biopsychosocial Interaction (5 th ed.)*. New Jersey John Wiley & Sons, Inc. Chapter 11 pp 291-318.

[11] Rolke R, Baron R, Maier C, Tölle T R, Treede R D, Beyer A, Binder A, Birbaumer n, Birklein F, Bötefür I C, Braune S, Flor H, Huge V, Klug R, Landwehrmeyer G B, Magerl W, Maihöfner C, Rolko C, Schaub C, Scherens A, Sprenger T, Valet M and Wasserka B 2006 *Quantitative Sensory Testing in the German Research Network on Neuropathic Pian (DPNS): Standardized Protocol and Reference Values* Pain 123 pp 231-243.

[12] Watson A, Deredy W E, Iannetti G D, Lloyd D, Tracey I, Vogt B A, Nadeau V and Jones A K P 2009 *Placebo Conditioning and Placebo Analgesia Modulate a Common Brain Network During Pain Anticipation and Perception*. Pain 145 pp 24 - 30.

[13] Michael N H, Otti A, Laeer L, Wohlschlaeger A, Zimmer C, Lahmann C, Toelle T and Guendel H 2010 *Aftermath of Sexual Abuse History on Adult Patient Suffering From Chronic Functional Pain Syndromes: An fMRI Pilot Study* Journal of Psychosomatic Research 68 pp 483-487.

[14] Newton-John T R O 2002 In a Sarafino E D 2006 *Health Psychology: Biopsychosocial Interaction (5 th ed.)*. New Jersey John Wiley & Sons, Inc.

[15] Kerns R D, and Weiss L H 1994 In a Sarafino E D 2006 *Health Psychology: Biopsychosocial Interaction (5 th ed.)*. New Jersey John Wiley & Sons, Inc.

[16] Silani G, Lamm C, Ruff C C, Singer T 2013 *Right Supramarginal Gyrus Is Crucial to Overcome Emotional Egocentricity Bias in Social Judgments* Journal of Neuroscience 33(39) pp 66 - 76

Effect of different thickness of material filter on Tc-99m spectra and performance parameters of gamma camera

A Nazifah[1], S Norhanna[1], S I Shah[1] and A Zakaria[2]

[1] International Islamic University Malaysia, Kuantan, 25200, Pahang, Malaysia
[2] University Sains Malaysia, Kubang Kerian, 16150, Kelantan, Malaysia

E-mail: nazifah@unisza.edu.my

Abstract: This study aimed to investigate the effects of material filter technique on Tc-99m spectra and performance parameters of Philip ADAC forte dual head gamma camera. Thickness of material filter was selected on the basis of percentage attenuation of various gamma ray energies by different thicknesses of zinc material. A cylindrical source tank of NEMA single photon emission computed tomography (SPECT) Triple Line Source Phantom filled with water and Tc-99m radionuclide injected was used for spectra, uniformity and sensitivity measurements. Vinyl plastic tube was used as a line source for spatial resolution. Images for uniformity were reconstructed by filtered back projection method. Butterworth filter of order 5 and cut off frequency 0.35 cycles/cm was selected. Chang's attenuation correction method was applied by selecting 0.13/cm linear attenuation coefficient. Count rate was decreased with material filter from the compton region of Tc-99m energy spectrum, also from the photopeak region. Spatial resolution was improved. However, uniformity of tomographic image was equivocal, and system volume sensitivity was reduced by material filter. Material filter improved system's spatial resolution. Therefore, the technique may be used for phantom studies to improve the image quality.

1. Introduction

Nuclear medicine imaging is considered as an important modality amongst the other medical imaging techniques. It can provide physiological information at the molecular and cellular level that leads to the determination of health conditions. It can be categorized into, planar and tomography. Tomography is most commonly performed because it can provide the 3D information about physiologic functions of organ e.g, perfusion, and metabolism [1] and high diagnostic accuracy [2,3] which depends on the ability of the imaging device to detect differences in the uptake of a radiopharmaceutical in a lesion and its surroundings. The quality and details of images obtained by gamma cameras are affected by several parameters, such as spatial resolution, sensitivity and uniformity [4]. Further, NaI (Tl) detector of the gamma camera is unable to discriminate all scattered gamma photons due to poor energy resolution. That causes poor quality image and inaccuracy in the quantification of radiopharmaceutical uptake [5–7]. Thus scatter correction is important for absolute quantification of single photon emission computed tomography (SPECT) data reported [8–10]. As stated by Hutton [11] there are many scatter correction techniques developed in last 3 decades, but they have some drawbacks for routine implementation. Another technique which reduces some fraction of scatter by using a material filter applied in planar imaging by Pillay [12] which is an alloy filter consisting of Pb, Zn and Sn for patient studies and they discovered improvement in image contrast and suggested this method may be used in SPECT. Study conducted by Pollard [13] for imaging therapeutic doses of -131, attaching physical filter of lead sheets (1.6mm - 6.4mm). In PET scanning, lead filters were used in order to reduce the number of scattered events [14]. Further, the effects of material filters, i.e., copper and aluminum to reduce scattered gamma photons in SPECT using phantom were studied by Shah [15]. Based upon the above findings, Zn material of different thicknesses was used as a material filter to investigate the effects on imge quality in SPECT.

2. Materials and methods

2.1. Calculation for selection of material filter.

Using equation 1 calculations were carried out for zinc (0.1 - 1 mm thicknesses) to determine the attenuation fraction of different gamma photon energies.

$$I = I_0 e^{-\mu x} \quad (1)$$

Where I_o is the intensity of the gamma photon beam before attenuation by the material, I is the intensity of the transmission gamma photons after passing through the material and μ is the linear attenuation coefficient (cm^{-1}) of material at a given energy of gamma photon, x is the thickness of the material. Linear attenuation coefficient (μ) values were taken from [16] for gamma photon energies ranging from 80 – 200 keV.

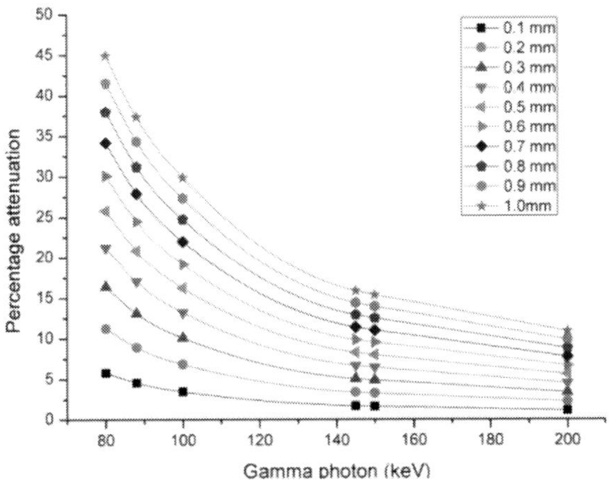

Figure 1. Percentage attenuation of gamma photon energies by various thickness of zinc.

Based on theoretical values, i.e., percentage difference of gamma photon energies attenuated by various thicknesses of material versus energy has been plotted as shown in Figure 1. As expected, the percentage of gamma photon attenuation decreases with increasing energy and increases with increasing thickness of material. However, the relative percentage attenuation of zinc 0.1 mm thick material is higher within the range of (92-139 keV) compared to 140 keV photons. But, due to the nonavailibility of 0.1 mm thickness, material filters of 0.198 mm and 0.4 mm thickness have been designed.

2.2. Planar data acquisition

2.2.1. Measurement of energy spectra of Tc 99m. Cylindrical source tank of NEMA SPECT Triple Line Source Phantom (20.2 cm inner diameter and inner height 20 cm) was filled with water and Tc-99m (15 mCi) was injected. Mineral water was used to make sure that there are no dust particles. After that, small amount of ink was added to provide a visual indication of the distribution of the activity. Cylindrical phantom was shaken gently for uniform distribution of radioactivity. Cylindrical tank was placed on the patient couch closer to the low energy high resolution *(LEHR)* collimator face in the central FOV of a gamma camera *(Philip ADAC Forte dual head)*. Matrix size was used is 64 x 64 x 16. The energy window ranges, 54 – 160, 62-160, 70-160 keV until 156 – 160 keV. Duration of count rate measurement for every window without and with material filter was maintained at 5 seconds.

Scanning time and count rate for every reading in each energy window were recorded. Three readings for each window were recorded and corrected for decay.

2.2.2. Spatial resolution. To investigate the effect on spatial resolution of a material filter, a vinyl plastic tube having an inner diameter 0.8 mm and length 10 cm filled with a Tc-99m (1.36 mCi) was used as a line source. The acrylic plates (30 cm x 30 cm) of different thicknesses were employed as a uniform scattering medium (tissue equivalent) [17]. Data acquisition matrix size 64 x 64 x 16 was selected. 20% energy window centered at 140 keV was adjusted. Pixel size in the X-direction was calculated by dividing field of view (FOV) gamma camera 508 mm with number of pixels (64). The system spatial resolution was evaluated at a distance from 1 cm up to 10 cm in 1 cm increments collimator to source distance in air Figure 2 (a). Also with and without material filter figure 2 (b). Experimental setting was same as in the case of air but with an increment of scattering material thickness. Care was taken to make sure the position of line source to remain same every time while adding the scattering material. The time per image was maintained at 5 seconds. Line spread function (LSF) for every source to collimator distance was obtained from the line profiles. The gamma camera stores counts from a line source in a single view and count profile bell-shaped (Graph of counts per pixel vs distance along counts profile in pixels) was drawn perpendicularly to the line source at the same place for each data set. The count profiles of LSF was used to determine the full width at half maximum (FWHM) and full width at tenth (FWTM).

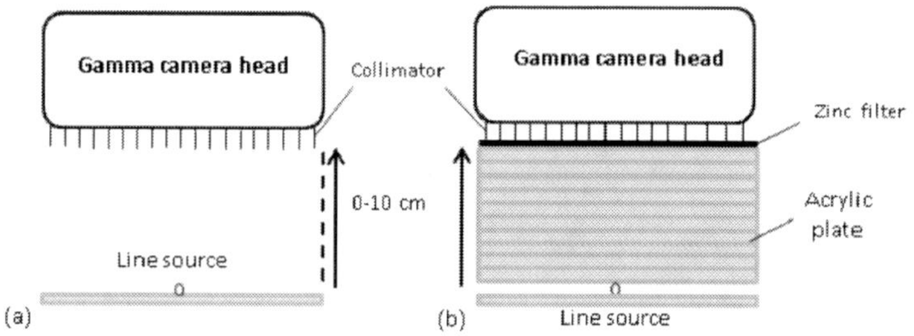

Figure 2. Arrangement for measurement of spatial resolution (a) without material filters (b) with scattering medium.

2.3. Tomography (SPECT) data acquisition

2.3.1. Uniformity. For tomographic uniformity data without and with material filter (0.198 mm and 0.4 mm) a cylindrical source tank of NEMA SPECT Triple Line Source Phantom with volume (6,409.48 cm^3), containing a homogeneous solution of Tc-99m (7.15 mCi) was used. Data collection parameters were chosen, such as; matrix size 128 x 128 x 16, clockwise rotation 360^0. Ninenty projections, 21 seconds per projection with 15 million total counts as recommended by [18], rotation radius of gamma camera head was fixed at 16.9 cm parallel to the axis of rotation. Energy window was 20% centered at 140 keV. SPECT images from the phantom data were reconstructed by filtered backprojection method. Butterworth filter of order 5, cut off frequency 0.35 cycles/cm was selected. Chang's attenuattion correction was applied by selecting 0.13/cm linear attenuation coefficient.

2.3.2. System volume sensitivity. Homogeneous solution of Tc 99m (16.74 mCi), 128 projections with 100, 000 counts per projection followed the same procedure as by [19]. The rest of the data collection parameters were used same as described in section 2.2.1.1. Activity and time to start and finish was recorded as shown in Table 1. System volume sensitivity was calculated using equation (17-15) and also volume sensitivity per axial, by dividing with diameter of cylindrical source tank [17].

Table 1. Activity, total scans time and total number of counts.

	Activity at scanning time (mCi)	Total time of scanning (hours)	Number of counts
Without filter	14.72	1.12	86, 682.08
Material 0.198 mm	13.97	1.55	82, 419.23
Material 0.4 mm	13.26	2.10	78, 063.93

3. Results and Discussion

3.1. Tc- 99m energy spectra.
Figure 2 shows that energy spectra of Tc 99m without and with material filter (0.198 mm and 0.4 mm) that was plotted by imaging cylindrical tank. As some scatter is present into photopeak region. This results maybe explained by the fact that gamma rays that have undergone large-angle scatters in the cylindrical tank and emerge with lower energy. This is significant problem that is encountered in clinical studies because scattered photons emit from the patient. With material filter, it should be noted that not only count rate in compton region is decreased but also in the photopeak region. That means it were reduce scatter portion in photopeak region. And this is seen clearly at compton and photopeak regions from the table 2. Research conducted by Kojima [20] show that in 125-139 keV region 75-80% are present and also stated by Hutton [11] about 30-40% scatter counts are in photopeak region. Results shows that when material filter 0.4 mm was applied it reduces more than double of counts as compared to material filter 0.198 mm for both of regions. So the most ideal filter here might be material filter 0.198 mm because reduces less count rate at 140 keV.

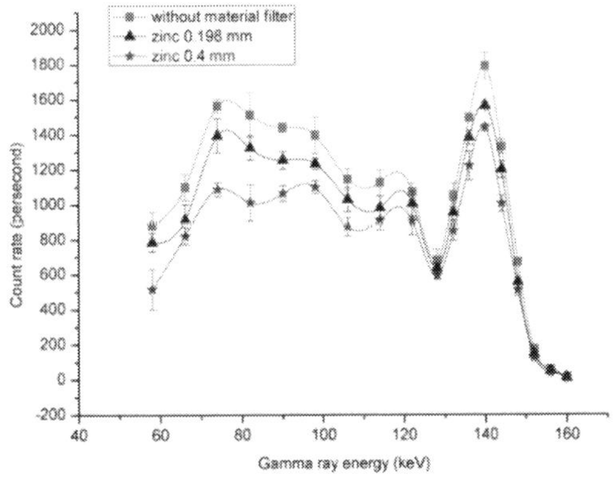

Figure 3. Energy spectra of Tc-99m without and with material filter (0.19 mm and 0.4 mm).

Table 2. Percentage reduction in count rate with material filter (0.198 mm and 0.4 mm).

Material	Compton region		Photopeak region	
thickness(mm)	92-125 keV	126-139keV	140 keV	141-154 keV
0.198	10.99%	8.75%	13.14%	12.87%
0.4	22.44%	18.49%	21.28%	26.45%

3.2. Spatial Resolution
From the Figure 4 and 5 the presence of scattering material degrade spatial resolution as compared to air. In this study, when material filter 0.198 mm was applied improvement in FWHM and FWTM below 7 and 5 cm source to collimator distance was obtained, respectively. Another finding was that FWHM and FWTM were degraded when material filter 0.4 mm was applied as compared to 0.198 mm. This is because of the increased thickness of material filter which has most probaly absorbed more unscattered events and also generated more scattered gamma photons. This finding is in agreement with ALehyani [21] statement that with the increase in source to collimator distance spatial resolution of the gamma camera was degraded.

3.3. Uniformity
Degree of uniformity over each transverse image of cylindrical phantom was evaluated by drawing count profile curves along the X-axis through the centre. Transverse image number twenty three was selected because it was more uniform as compared to other images. From the Figure 5 below we can see the count profile curve along the X- axis of the cylindrical phantom. However, no difference was observed in the count profile with both thicknesses of material filter (0.198 mm and 0.4 mm) as compared to without material filter. For the future research it is suggested for increase in the amount of Tc-99m.

3.4. System volume sensitivity

Table 3 shows that when a material filter 0.198 mm was applied, 5.69% reduction as compared to without material filter and that reduction increased to 13% when material filter 0.4 mm. Furthermore, results are similar to volume sensitivity per axial. These results might be due to absorption count through a material filter that have reduction in system volume sensitivity and also for volume sensitivity per axial.

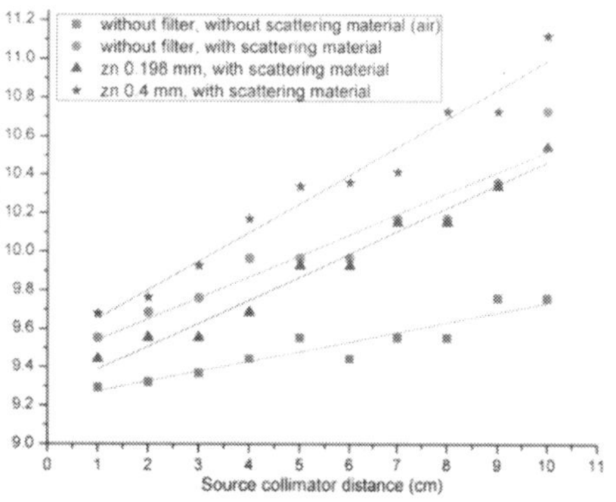

Figure 4. FWHM in air and in scattering medium (without and with material filter).

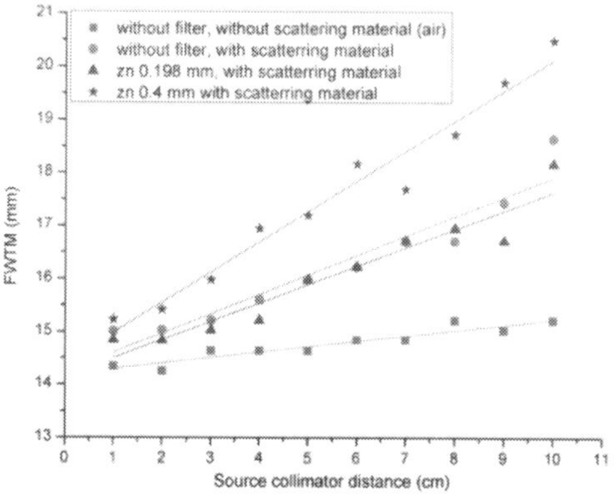

Figure 5. FWTM in air and in scattering medium (without and with material filter).

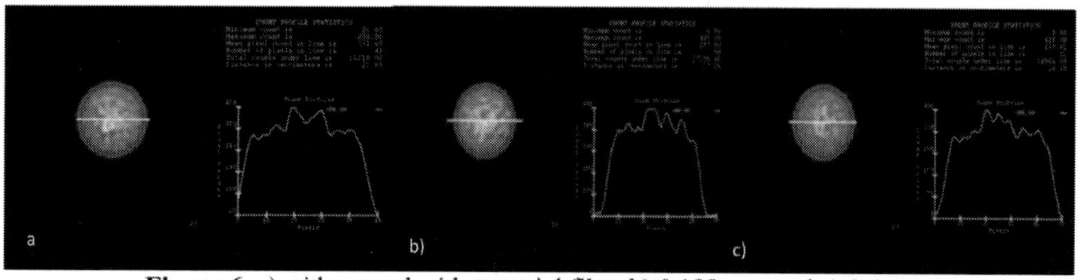

Figure 6. a) without and with material filter b) 0.198 mm and c) 0.4 mm.

Table 3. Concentration of radionuclide, total time acquisition, number of counts, system volume sensitivity and volume sensitivity per axial.

(mm)	A Concentration of radionuclide (Bq/mL)	T Time required to complete acquisition (sec)	N number of counts recorded	SVS (cps/Bg/mL)	Volume sensitivity per axial
Without material filter	84,974.13	1260	13215302	0.123	6.09×10^{-3}
Material 0.198	80,817.79	1320	12333928	0.116	5.74×10^{-3}
Material 0.4	76,545.99	1380	11279588	0.107	5.30×10^{-3}

4.Conclusion

With the use of material filter, it shows the count rate is decreased from compton and photopeak region of Tc-99m spectrum. Material filter 0.198 mm thickness improved spatial resolution, however, uniformity of tomographic images is equivocal, and system volume sensitivity has shown reduction as compared to without material filter. Therefore, based on spatial resolution results the material filter may be applied for improvement in image quality in Tc-99m planar and tomography. This technique reduce the scattered gamma photons from the raw data before image reconstruction. In terms of clinical implementation, it is easy and cost effective.

Acknowledgment

Authors are grateful to the Department of Nuclear Medicine, Oncology and Radiotheraphy Hospital Universiti Sains Malaysia, Kubang Kerian, Kelantan for the use of Gamma Camera and other facilities to carry out this research work. Also thanks to School of Health Sciences Universiti Sains Malaysia for other support.

References

[1] Schaefer W M, Nowak B, Kaiser H, Koch K, Block S, Dahl J 2003 Comparison of Microsphere-Equivalent Blood Flow (^{15}O-Water PET) and Realtive Perfusion (^{99m}Tc-Tetrofosmin SPECT) in Myocardium Showing Metabolism – Perfusion Mismatch *J Nuc Med* **44** 33–9

[2] uz Zaman M 2007 99mTc-EC-deoxyglucose-- poor man's 18F-FDG: what will be the future of PET in molecular imaging? *Eur J Nucl Med Mol Imaging* **34** 429 doi:10.1007/s00259-006-02998 8

[3] Tamahi N, Tadamura E, Kudoh T 1997 Recent advances in nuclear cardiology in the study of coronary artery disease *Ann Nucl Med* **11** 55–66

[4] Lyra M, Ploussi A 2011 Filtering in SPECT Image Reconstruction *Int J Biomed Imaging* 1-14 doi: 10.1155/2011/693795

[5] Chuanyong B, Babla H, Conwell R 2010 Emission-Based Scatter Correction in SPECT Imaging *Tsinghua Sci Tehnol* **15** 1–10

[6] Hesse B, Tägil K, Cuocolo A, Anagnostopoulos C, Bardiés M, Bax J, et al 2005 EANM/ESC procedural guidelines for myocardial perfusion imaging in nuclear cardiology *Eur J Nucl Med Mol Imaging* **32** 855–97 doi: 10.1007/s00259-005-1779-y

[7] de Vries D J, King M A, Soares E J, Tsui B M W, Metz C E 1999 Effects of Scatter Subtraction on Detection and Quantitation in Hepatic SPECT *J Nucl Med* **40** 1011–1023

[8] Xiao J, de Wit T C, Staelens S G, Beekman F J 2006 Evaluation of 3D Monte Carlo – Based Scatter Correction for 99m Tc Cardiac Perfusion SPECT *J Nucl Med* **47** 1662–1669

[9] Celler A, IEEE M, Axent D, Toganet D, El-Khatib J 2000 Investigation of Scatter in SPECT Transmission Studies *IEEE Trans Nucl Sci* **47** 1251–1257

[10] Ljungberg M, King M A, Hademenos G J, Strand S 1993 Comparison of Four Scatter Correction Methods Using Monte Carlo Simulated Source Distributions *J Nucl Med* **35** 143–151

[11] Hutton B F, Buvat I, Beekman J F 2011 Review and current status of SPECT scatter correction *Phys Med Biol* **56** R85–R112 doi: 10.1088/0031-9155/56/14/R01

[12] Pillay M, Shapiro B, Cox P H 1986 The effect of an alloy filter on gamma camera images. Eur *J Nucl Med* **12** 293–5

[13] Pollard K R, Bice A N, Eary J F, Durack L D, Lewellen T K 1992. A Method for Imaging Therapeutic Doses of Iodine- 13 1 with a Clinical Gamma Camera *J Nucl Med* **33** 771–6

[14] Spinks T J, Shah S I 1993 Effect of lead filters on the performance of a Neuro-PET Tomography operated without septa *IEEE Trans Nucl Sci* **40** 1087–1091

[15] Shah S I, Zakaria A, Yee F S 2004 Effect of Unconventional Filters on Uniformty, Sensitivity and Linearity of an Imaging System in SPECT *Australas Phys Eng Sc* 27

[16] Hubbell J H, Seltzer S M 1996 NIST: X-Ray Mass Attenuation Coefficients Radiat Biomol Phys Div PML, NIST [cited 2014 Apr 14]; Available from: http://www.nist.gov/pml/data/xraycoef/

[17] International Commission on Radiation Units and Measurements (ICRU) [cited 2014 April Available from: http://www.icru.org/home/reports/tissue-substitutes-in-radiation-dosimetry-and-measurement-report-44

[18] Assurance Q, Systems S. IAEA Human Health Series No.6 Quality Assurance for SPECT Systems 6

[19] Cherry S R, Sorenson J A, Mutsumasa T, Phelps M E 2012 Physics in Nuclear Medicine. *Fourth edi. Elsevier Saunders*

[20] Kojima A, Matsumoto M, Mutsumasa T, Takahashi M 1991 Eksperimental analysis of scattered photons in Tc-99m imaging with a gamma camera *Ann Nucl Med* **5** 139–44

[21] ALehyani S H A 2009 Application of single photon emission computed tomography (SPECT) parameters for bone scintigraphy *J King Saud Univ* **21** 107-117

Estimation of patient radiation dose from whole body ^{18}F-FDG PET/CT examination in cancer imaging: a preliminary study

M H Mahmud[1,2], A J Nordin[2], F F Ahmad Saad[2] and A Z Fattah Azman[3]

[1] Faculty of Medicine and Health Sciences, Universiti Putra Malaysia 43400 UPM Serdang, Selangor, Malaysia

[2] Centre for Diagnostic Nuclear Imaging, Universiti Putra Malaysia 43400 UPM Serdang, Selangor, Malaysia

[3] Department of Community Health, Faculty of Medicine and Health Sciences, Universiti Putra Malaysia 43400 UPM Serdang, Selangor, Malaysia

E-mail: hafizimahmud@yahoo.com

Abstract. This study aims to estimate the radiation effective dose resulting from whole body fluorine-18 flourodeoxyglucose Positron Emission Tomography (^{18}F-FDG PET) scanning as compared to conservative Computed Tomography (CT) techniques in evaluating oncology patients. We reviewed 19 oncology patients who underwent ^{18}F-FDG PET/CT at our centre for cancer staging. Internal and external doses were estimated using radioactivity of injected FDG and volume CT Dose Index ($CTDI_{vol}$), respectively with employment of the published and modified dose coefficients. The median differences of dose among the conservative CT and PET protocols were determined using Kruskal Wallis test with $p < 0.05$ considered as significant. The median (interquartile range, IQR) effective doses of non-contrasted CT, contrasted CT and PET scanning protocols were 7.50 (9.35) mSv, 9.76 (3.67) mSv and 6.30 (1.20) mSv, respectively, resulting in the total dose of 21.46 (8.58) mSv. Statistically significant difference was observed in the median effective dose between the three protocols ($p < 0.01$). The effective doses of whole body ^{18}F-FDG PET technique may be effective the lowest amongst the conventional CT imaging techniques.

1. Introduction

The integration of Positron Emission Tomography (PET) and Computed Tomography (CT) scanners or the PET/CT provides a co-registration of morphological and functional information in a single setting. The technology of CT-based attenuation correction for the PET acquisition data is preferred to the out-phased conventional gamma ray source (such as germanium-68) and has greatly reduced the scanning time in a PET/CT study [1] and improved the quality of the corrected PET scans [2-4]. This integrated modality is becoming an important tool nowadays for clinical investigation with increased clinical use, expanding from oncologic diagnosis to other clinical indications, including infection, inflammation, cardiac study and neurology. The PET/CT plays an important role in clinical applications and yields as much imaging information as possible to derive an accurate diagnosis and to indicate the most appropriate treatment that the patient can receive. However, concern over the total radiation dose to patients in the PET/CT study is a matter of great concern among the clinicians.

To the best of our knowledge, there is limited study addressing the radiation dose of patients undergoing whole body fluorine-18 fluorodeoxyglucose (^{18}F-FDG) PET/CT examinations in Malaysian population, in particular. Therefore, this study is aimed to estimate the radiation effective dose resulting from whole body ^{18}F-FDG PET scanning as compared to conservative CT techniques in evaluating oncology Malaysian patients.

2. Materials and methods

2.1 Patients

This retrospective study included the analysis of integrated ^{18}F-FDG PET/CT images from 19 patients (10 men and 9 women; mean age 53.2 ± 16.8 years; ranging between 20 and 75 years) who were referred for whole body ^{18}F-FDG PET/CT imaging for cancer staging at the Centre for Diagnostic Nuclear Imaging of Universiti Putra Malaysia, Malaysia. This study was granted ethical approval by the institutional review committee.

2.2 Patient preparation

All the patients fasted for at least 6 hours before the scanning session and only oral hydration with glucose-free water was allowed. The fasting blood glucose level was recorded for all patients. Oral gastrografin solution (sodium meglumine diatrizoate; BerliMed S.A., Madrid, Spain) was given to patients in a three part dilution before the intravenous (IV) administration of a radiopharmaceutical agent and immediately before scanning. Mean 341.3 MBq (within the range of 293.8 - 397.8 megabecquerels) of ^{18}F-FDG was injected intravenously. All the patients were put to rest in a special uptake room for an average of 80.4 min (within the range of 30 - 282 min) and emptied their bladder before undergoing the PET/CT examination.

2.3 PET/CT imaging protocol

Image acquisition was performed using an integrated Siemens Biograph 64 TruePoint PET/CT system (Siemens Medical Solutions, Erlangen, Germany) consisting of a PET scanner with lutetium oxyorthosilicate (LSO) crystals detector and a 64-multi detector CT scanner (MDCT). A scout view was performed in the cranio-caudal direction to plan the study, followed by a non-contrast enhanced CT (NECT) protocol in caudo-cranial direction for the purpose of anatomical localization and attenuation correction. Subsequently, a contrast-enhanced CT (CECT) or known as high dose CT protocol with IV injection of non-ionic contrast, iohexol (Omnipaque 350 mgI/mL, GE Healthcare, Shanghai, China) 84.1 mL (mean; within the range of 50 - 110 mL) was performed, using a dual head automatic pressure injector (Mallinckrodt, Missouri, USA) with a flow rate of 2.5 mL/s and was followed by a 20 mL saline flush. CECT acquisition started in the caudo-cranial direction. A PET scan was acquired consecutively at 3 min per bed position using a three-dimensional acquisition mode. The total duration of the PET/CT examination was about 20 min, with approximately 1 min to complete the CT scans and about 19 min to acquire the PET emission data. The latter was acquired in a caudo-cranial direction. The PET images were then reconstructed by using a TrueX reconstruction algorithm. The CT datasets were employed for attenuation correction with the same set of PET images.

2.4 Estimation of internal dose

Absorbed dose , D_T to a tissue or organ, T resulting from the IV administration of an activity, A of ^{18}F-FDG was computed by means of dose coefficients Γ_T^{FDG} as recommended by the International Commission on Radiological Protection (ICRP) in Publication 106 [5] for a variety of organs and tissues of the adult hermaphrodite MIRD phantom, which is $D_T = A\,.\,\Gamma_T^{FDG}$. Whole body effective dose coefficient values were calculated using the acquired data as reported by Brix et al. [6] and Huang et al. [7] with some modifications on these data through the use of a tissue weighting factor, W_T provided by ICRP Publication 103 [8].

Using the published and modified whole body effective dose coefficient values, the average effective dose for whole body FDG PET was estimated as follows [6] :

$$E = \sum_T w_T \,.\, D_T = A \,.\, \sum_T w_T \,.\, \Gamma_T^{FDG} = A \,.\, \Gamma_E^{FDG} \tag{1}$$

2.5 Estimation of external dose

To estimate the radiation exposure of patients resulting from the acquisition of CT scans, volume CT Dose Index ($CTDI_{vol}$) was used; it was directly obtained from the display screen of the operator console of the CT workstation. The organ specific dose coefficient Γ_T^{CT} values for organ dose calculation were also adapted from Brix and his co-workers [6]. With some modifications through the use of a tissue weighting factor, W_T provided by the ICRP Publication 103 [7], the whole body effective dose coefficients Γ_E^{CT} value was calculated using the following formula [6]:

$$\Gamma_E^{CT} = \sum_T w_T \,.\, \Gamma_T^{CT} \tag{2}$$

Using both whole body effective dose coefficients, as adapted from the study by Brix et al. [6] and the modification method, the average effective dose for whole body CT was estimated by equation [9]:

$$E = \Gamma_E^{CT} \,.\, CTDI_{vol} \tag{3}$$

2.6 Statistical analysis

The results were presented as a median (interquartile range, IQR). The Kruskal-Wallis test was carried out to determine the difference of the median effective dose between each of the stand-alone scan protocol, including NECT, CECT and PET. All the hypothesis tests were two-sided with a significant level of 0.05. The Statistical Package for the Social Sciences program for Windows 21.0 (SPSS 21) (IBM Corp, Somers, New York) was used for the statistical analysis.

3. Results and discussion

The median whole body effective doses for NECT and CECT scans were 7.50 (IQR 9.35) mSv (ranging from 0.61 to 43.80 mSv) and 9.76 (IQR 3.67) mSv (ranging from 7.01 to 27.56 mSv), respectively. The median whole body effective doses from the PET were 6.30 (IQR 1.20) mSv (ranging from 5.45 to 7.38 mSv) whereby 21.46 (IQR 8.58) mSv (ranging from 15.00 to 77.38 mSv) contributed to the total whole body effective dose for PET/CT scan. The data on the patient effective dose from the stand-alone scan and whole body PET/CT scan are shown in Table 1.

This study demonstrated that CT examination took the major role in contributing to the total effective dose of PET/CT imaging, corresponding to approximately 80.43%. This finding is in agreement with a study by Huang et al. [7], in which up to 81% of the total PET/CT effective dose was attributable to the CT doses. Huang and his colleagues specified that diagnostic CT scanning is a standard practice for PET/CT imaging in their institution, which is similar to the practice in the present study. In fact, their PET/CT system (64-detector CT system) and some CT scanning parameters (kVp and scanning coverage area) were also similar to those in the present study as well, making the data comparable.

In this particular study, the average FDG radioactivity of 341.3 ± 37.6 MBq was administered for the patients, this level of radioactivity being derived from their body weight. This amount is greater than the value in a study by Khamwan et al. [10], even though the physical characteristics of their Thai patients seem not to have been greatly different from those of the Malaysian patients in the present study. With this higher value, it was expected that the effective dose of PET scan from this study would be increased. Nevertheless, the effective dose of PET resulting from this study is still in accordance with the results in the study by Brix et al. [6] with a range of 5.7 to 7.0 mSv in the effective dose of PET reported in their study. Moreover, the total PET/CT effective dose for the Thai patients seems to have been lower than the effective dose in the present study. The present NECT study was acquired with the mid range exposure which has given higher dose in comparison to low dose CT. A higher tube current exposure time product (mAs) is required to compensate the increase in density in the contrasted tissues as a result of contrast media administration. The additional CT dose contributing from CECT results to increase the total dose as compared to the NECT alone. In fact, only the NECT protocol was performed in all CT scans in the PET/CT practice of the Thai study; a much lower dose is expected from such CT scans than in the integrated protocols (NECT and CECT protocols) that the present study records. However, when comparing the effective dose of NECT with that given to the Thai patients, a much lower effective dose of 7.50 mSv was demonstrated in the present study than for the Thai patients (14.45 mSv).

Table 1. The effective dose of the stand-alone and integrated whole body PET/CT scanning protocols.

Patient no.	Effective dose (mSv)			
	Stand-alone			Integrated PET/CT scan
	NECT[a]	CECT[b]	PET[c]	NECT+CECT+PET
1	0.75	12.08	7.31	20.14
2	9.70	10.06	6.70	26.46
3	0.65	9.34	6.77	16.76
4	0.73	12.08	7.23	20.04
5	0.62	8.24	6.56	15.42
6	0.75	12.08	6.82	19.65
7	0.65	8.41	7.38	16.44
8	0.62	9.34	7.33	17.29
9	0.61	8.36	6.03	15.00
10	7.50	7.26	6.70	21.46
11	7.23	7.01	5.81	20.05
12	43.80	27.56	6.02	77.38
13	10.00	9.76	5.64	25.40
14	9.88	9.21	5.62	24.71
15	10.25	10.00	5.62	25.87
16	9.09	8.54	5.49	23.12
17	14.76	13.72	5.45	33.93
18	12.75	12.39	6.30	31.44
19	9.70	10.36	5.50	25.56

[a] Effective dose acquired from non-contrast enhanced CT scan
[b] Effective dose acquired from contrast-enhanced CT scan
[c] Effective dose acquired from PET scan

The median effective dose for patients undergoing whole body ^{18}F-FDG PET/CT imaging in our study was 21.46 (8.58) mSv, which is less than is specified in the previous reports by Brix et al. [6] of 25.0 mSv and by Chawla et al. [11] of 24.8 mSv. Indeed, this result is within the range of the mean effective dose values for male and female patients as reported by Huang et al. [7], with effective doses ranging from 13.65 to 32.18 mSv and 13.45 to 31.91 mSv, respectively. These findings demonstrated that the effective doses resulting from the current protocols employed in this study were still lower than in the previously reported studies concerning the radiation dose to patients in a PET/CT imaging study. Therefore, it is justifiable to note that the radiation dose resulting from the PET/CT scan has to be tailored to the needs of the study and the impact of doing so should outweigh the radiation effect.

Variation to the basic CT practice has been employed to optimize the diagnostic capability of both CT and PET systems during an integrated PET/CT study. Low dose CT scan was not employed in our CT scanning protocol rather than NECT and diagnostic CT scans. Different approach of CT practice affects the resulting CT radiation dose in different way. Even though the utilization of IV contrast enhanced CT has been reported to cause overestimation on the standardized uptake value (SUV) of attenuation corrected PET, which leading to contrast-induced artefacts on PET images [12], however, several evidences have revealed that the effects were statistically and clinically insignificant [13-15]. We optimized the advantage of diagnostic CT in our practice in cases where lesions are non PET-avid. The technique is capable in characterizing tumour physiology with improved visual delineation of tumour margins [16]. As the CT detail information can be optimized and utilized in patient clinical management, the diagnostic CT is being used as a protocol of choice for selected malignant diseases in our PET/CT study.

Our study was limited and had a small number of patients. Hence, to determine better the effective dose in a PET/CT study, further investigation with employment of various CT protocols and a larger group of samples presenting various indications for PET/CT study is warranted. A larger study would ensure a more significant result as there is more variation in the numerical data analysed. Variation in the technique such as low dosed CT would be carried out to test the effect of the image quality and hence would potentially be used as an alternative to the routine NECT technique. Moreover, the method of dose estimation is another limitation in this study. The use of special software for better and more accurate estimation of the effective dose of PET such as OLINDA [17] and effective dose of CT such as CT-Expo [18] and ImPACT [19] are recommended. Since these software programs were not available in our study, the estimation of an effective dose could only be derived from calculation, employing the available dosimetric descriptors including $CTDI_{vol}$ and the activity of the administered FDG.

4. Conclusion

The effective doses of whole body ^{18}F-FDG PET scan may be effective the lowest amongst the conventional CT imaging techniques. The median effective dose for patients undergoing whole body ^{18}F-FDG PET/CT examination at the Centre for Diagnostic Nuclear Imaging of Universiti Putra Malaysia was 21.46 mSv. The radiation exposure of patients during PET/CT study must be justified by clinical outcome throughout the diagnostic process.

Acknowledgements

The authors would like to thank the radiographers of the Centre for Diagnostic Nuclear Imaging of Universiti Putra Malaysia for their assistance in retrieving the PET/CT images and all the staff members of the centre for their kind support in this study. A special word of thanks goes as well to the Universiti Putra Malaysia for funding this work under Research University Grant Scheme (grant number 9331800).

References

[1] Blodgett TM, Meltzer CC and Townsend DW 2007 *Radiology* **242** 360-385
[2] Townsend DW, Carney JP, Yap JT and Hall NC 2004 *J. Nucl. Med.* **45** 4S-14S
[3] Nakamoto Y, Osman M, Cohade C, Marshall LT, Links JM, Kohlmyer S and Wahl RL 2002 *J. Nucl. Med.* **43** 1137-43
[4] Wu TH, Huang YH, Lee JJS, Wang SC, Su CT, Chen LK and Chu TC 2004. *Eur. J. Nucl. Med. Mol. Imaging* **31** 38-43
[5] 2008. *Radiation dose to patients from radiopharmaceuticals. Addendum 3 to ICRP Publication 53.* Ontario: International Commission on Radiological Protection. ICRP publication 106
[6] Brix G, Lechel U, Glatting G, Ziegler SI, Munzing W, Muller SP and Beyer T 2005 *J. Nucl. Med.* **46** 608-13
[7] Huang B, Law M W-M and Khong P-L 2009 *Radiology* **251** 166-74
[8] 2007. *The 2007 recommendations of the International Commission on Radiological Protection.* Ontario: International Commission on Radiological Protection. ICRP Publication 103
[9] Brix G and Beyer T 2005 *Nuklearmed.* **44** S51-7
[10] Khamwan K, Krisanachinda A and Pasawang P 2010 *Radiat. Prot. Dosim.* **141** 50-5
[11] Chawla SC, Federmna N, Zhang D, Nagata K, Nuthakki S, McNitt-Gray M and Boechat MI 2010 *Pediatr. Radiol.* **40** 681-6
[12] Mawlawi O, Erasmus JJ, Munden RF, Pan T, Knight AE, Macapinlac HA, Podoloff DA

and Chasen M 2006 *Am. J. Roentgenol.* **186** 308-19
[13] Hairil Rashmizal AR, Geso M., Noraini AR and Abdul Jalil N. 2010 *ANZ Nucl. Med.* **41** 19-23.
[14] Bunyaviroch T, Turkington TG, Wong TZ, Wilson JW, Colsher JG and Coleman R. 2008. *Mol. Imaging Biol.* **10** 107-13.
[15] Yau YY, Chan W, Tam YM, Vernon P, Wong S, Coel M and Chu SKF 2005 *J. Nucl. Med* **46** 283-91.
[16] Bisdas S, Baghi M, Smolarz A, Pihno NC, Lehnert T, Knecht R, Mck MG, Vogl TJ, Tuerky S and Koh TS 2007 *Invest. Radiol.* **42** 172-79
[17] Stabin MG, Sparks RB and Crowe E 2005 *J. Nucl. Med.* **46** 1023-27
[18] Stamm G and Hagel HD 2002 *Fortschr. Röntgenstr.* **174** 1570-76
[19] 2006. *ImPACT's CT dosimetry tool* [online]. London: ImPACT. Available from: http://www.impactscan.org/ctdosimetry.htm [Accessed 15 March 2014].

Effects of Fe as a physical filter on spectra of Technitium-99m, uniformity, system volume sensitivity and spatial resolution of Philip ADAC Forte dual-head gamma camera

N Sohaimi[1], N Abdullah[1], S I Shah[1] and A Zakaria[2]

[1] Department of Diagnostic Imaging and Radiotherapy, Kulliyyah of Allied Health Sciences, International Islamic University Malaysia, 25200 Kuantan, Pahang, Malaysia
[2] School of Health Science, Universiti Sains Malaysia, 16150 Kelantan, Malaysia

E-mail: norhanna82@yahoo.com

Abstract. Single photon emission computed tomography (SPECT) imaging inherits some limitations, i.e., due to scattered gamma photons which degrade spatial resolution causes poor image quality. This study attempts to reduce a fraction of scattered gamma photons before reaching gamma camera detector by using Fe sheet (0.35 mm and 0.40 mm) as a physical filter. Also investigate the effects on spectra of Tc-99m, spatial resolution, system volume sensitivity and uniformity. The thickness of Fe physical filter is selected on the basis of percentage attenuation calculations of different gamma ray energies by various thicknesses of material. Data were acquired using Philip ADAC forte dual-head gamma camera without and with physical filter with LEHR collimator installed. For spectra, uniformity and system volume sensitivity, a cylindrical source tank filled with water added with Tc-99m was scanned. Uniformity and system volume sensitivity images were reconstructed with FBP method by applying Butterworth filter of order 5, cut-off frequency 0.35 cycles/cm and Chang's attenuation correction method using 0.13 cm^{-1} linear attenuation coefficient. Spatial resolution study was done by scanning a line source (0.8 mm inner diameter) of Tc-99m at various source-to-collimator distances in air and in scattering medium without and with physical filter. A substantial reduction in count rate from Compton and photopeak regions of Tc-99m spectra with physical filter is recorded. Improvement in spatial resolution with physical filter up to 4 cm source-to-collimator distance is obtained. System volume sensitivity was reduced and no improvement in uniformity. These thicknesses of physical filter may be tested further by scanning different planar/SPECT phantoms in Tc-99m imaging.

1. Introduction

Single photon emission computed tomography (SPECT) is an imaging that involved a gamma camera which will detect the gamma rays that come out from the patient body part which injected with radiopharmaceutical. SPECT has a main problem of scatter and absorption of gamma photons in the object/patient to be scanned. The scatter gamma photons results a poor quality images and non-linearity in the quantitation of radioactivity uptake [2,3].

In SPECT system, scintillation detectors NaI(Tl) are used in order to detect the emitted gamma photons from the patient body. Compton scattered photons occur within the body of the patient. The presence of scattered gamma photons in the image data degrades the spatial resolution and uniformity in the image. Also, causes the inaccuracy in the quantification of radioactivity distribution.

The problem of Compton scattered photons has been handled by using energy discrimination among the others by narrowing down the energy window. But this method has limitations and unable to eliminate all scattered gamma photons from projection data because of the poor energy resolution of NaI(Tl). Therefore, the data collected in photopeak energy window comprises primary (unscattered) and scattered gamma photons. Figure 1 shows the simulated distribution of scattered gamma photons into the photopeak of Tc-99m with several scattering orders [10]. This phenomenon was reported by Kojima A et al. [5] which stated that the scattered photons are distributed over the whole symmetric photopeak window and 75% to 80% of these scattered photons are present within the energy window below 140 keV. Therefore, a method for the reduction of the influence of low energy (scattered) gamma photons on SPECT projection data by means of physical filter is used.

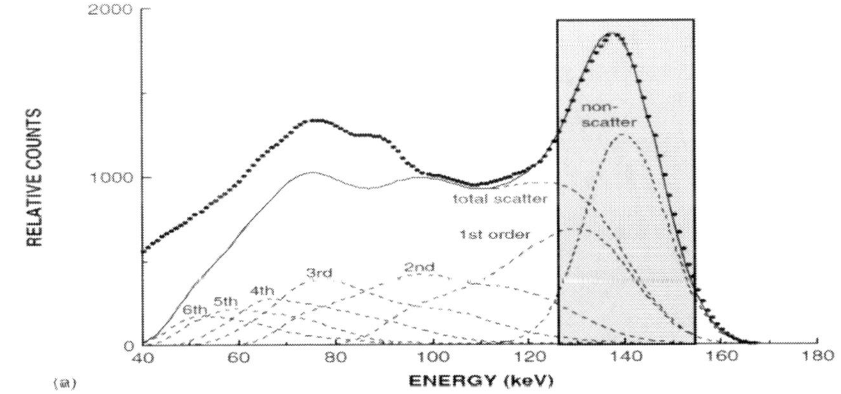

Figure 1. Energy spectra for a gamma emitting Tc-99m line source on the axis of a water-filled cylinder simulated using the Monte Carlo method adopted from Zaidi H 2006 [10].

The basic concept of physical filter is that it decreases the relative concentration of low energy (scattered) gamma photons by preferentially removing them before they can reach the surface of the detector as shown in Figure 2. Thus, the gamma camera may detect more unscattered photons and may provide more valuable information.

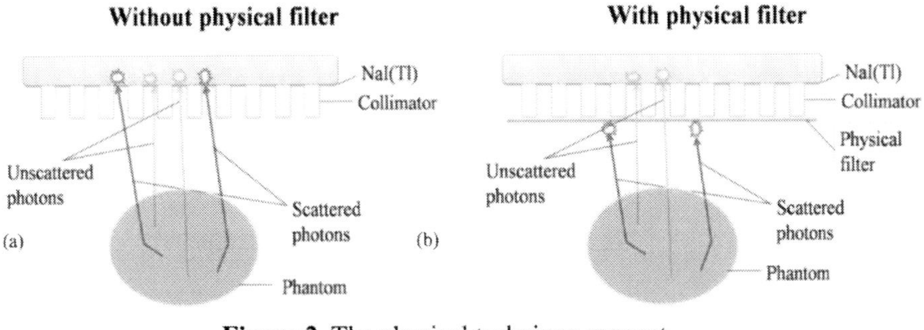

Figure 2. The physical technique concept.

Pillay M. et al [6] used an alloy filter that consist of Lead (Pb), Zinc (Zn), and Tin (Sn) in single photon planar clinical imaging and reported improvement in the image contrast. In 1990, Ficke and Ter-Pogossion [1] suggested that material filters might be advantageous in reducing low energy radiation originating in the field of view of the gamma camera. In their study, they analysed the spectra of F-18 without filter and with 0.017 inch and 0.034 inch thick Pb filters by using NaI(Tl) detector. Further, Spinks and Shah [9] investigated the effects of Pb filters (0.5 mm and 0.1 mm thick) on the overall performance of a Neuro-PET multi-ring (CTI/Siemens 953B) Tomograph and significant reduction in random rates and lower energy events in the measured energy spectrum were observed. A research done by Shah SI [7], using Tin (Sn) 0.25 mm thick sheet as a physical filter in Tc-99m SPECT. From the phantom study, the researcher found that there was reduction in low energy (scattered) gamma photons in the Tc-99m spectrum and improvement in the SPECT image contrast and SNR. In the year 2011, Shah SI et al [8] applied copper (Cu) material sheet 0.125 mm thick and aluminum (Al) 0.2 mm and image quality was marginally improved with physical filters. From the studies mentioned above, the application of physical filter is found to be significant and provide an easy way to reduce the scattered gamma photons from the image data.

The objective of this study is to investigate effects on spectra of Tc-99m, spatial resolution, system volume sensitivity and uniformity by using Fe physical filter.

2. Methodology

2.1. Selection of physical filter and thickness

Selection of physical filter depends on the type (characteristics) of materials such as linear attenuation coefficient and also the energy of the radioactive source used in the study. From the literature review, no one has used iron, Fe (Z=26) as a physical filter until recent. Thus, Fe was selected in this study. For the calculation purposes, a range of thicknesses were considered, between 0.05 mm to 0.50 mm. Percentage absorption of gamma photons of various energies was calculated by the equation:

$$\text{Percentage Absorption} = (1 - e^{-\mu x}) \times 100\% \qquad (1)$$

The linear attenuation coefficient, μ describes the fraction of a beam of x-rays or gamma rays that is absorbed or scattered per unit thickness of the absorber. The μ value was calculated from mass attenuation coefficient taken from Hubbell and Berger [4]. All percentage absorption values for material were tabulated in Table 1.

Table 1. Percentage absorption of Fe by various thicknesses for 126-139 keV and 140 keV of Tc-99m.

Thickness (mm)	Percentage absorption	
	126-139 keV	**140 keV**
0.05	9.23	8.34
0.10	17.81	16.16
0.15	24.79	22.59
0.20	32.54	29.79
0.25	38.53	35.42
0.30	44.24	40.84
0.35	47.52	44.01
0.40	53.00	49.29
0.45	55.98	52.22
0.50	61.88	57.96

From Table 1, the thicknesses were selected by the absorption percentage in terms of energy of gamma photons. 0.05 mm was the most appropriate thickness which absorbed 8.34% of 140 keV compared to 9.23% of 126-139 keV. The percentage difference between 140 keV and 126-139 keV is 8.9%. Unfortunately, the thickness (0.05 mm) was not available. Therefore, this study was done by applying 0.35 mm and 0.40 mm because of the availability of these thicknesses. The physical filter was designed sized 55 cm length and 44 cm width matching the size of gamma camera head. The surface of the physical filter was flat and due to the material fragility, it was laminated/coated with a thin laminating film.

2.2. Materials

Gamma camera system has been used is Philip ADAC Forte dual head attached with LEHR collimator and Pegasys software installed. In this study, Tc-99m was used as radioactive source because this source is easily available and commonly practiced in clinical studies.

2.3. Measurement of Tc-99m spectra without and with physical filter

The data for Tc-99m spectra were collected without and with physical filter by scanning a NEMA SPECT Triple Line Source Phantom™ (source tank only) filled with water and Tc-99m (15 mCi) distributed uniformly. The physical filter was mounted on the face of LEHR collimator as shown in Figure 3. Ten contiguous energy windows were adjusted over the entire spectrum of Tc-99m for recording the count rate. The spectra were plotted from the count rate obtained without and with physical filter versus the energy (keV). The count rates were corrected for the decay.

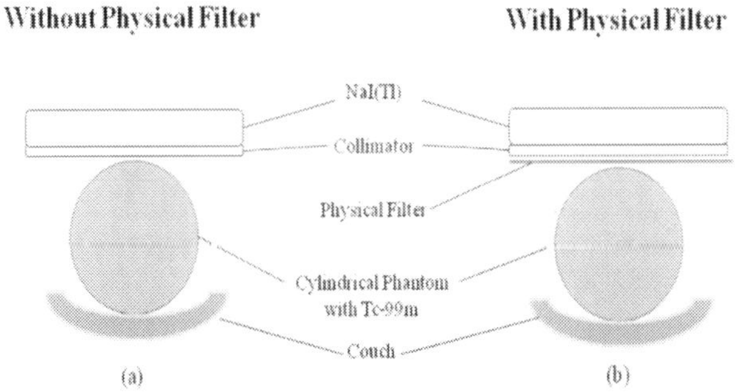

Figure 3. Experimental setup for Tc-99m spectra measurement (a) without and (b) with physical filter.

2.4. Measurement of spatial resolution without and with physical filter in air and scattering medium
The data were collected by scanning a line source (inner diameter: 0.8 mm) of Tc-99m (1.36 mCi) without and with physical filter within a 20% symmetrical energy window centred at the photopeak (140 keV). The physical filter was mounted on the face of LEHR collimator. The line source was scanned in air and scattering medium (acrylic plates) at various source-to-collimator distances ranging from 1 to 10 cm as shown in Figure 4.

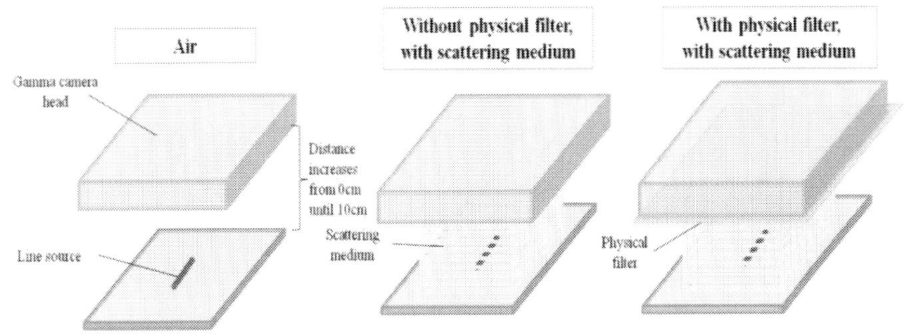

Figure 4. Experimental setup for measurement of spatial resolution without and with physical filter in air and in scattering medium.

2.5. Measurement of system volume sensitivity and uniformity without and with physical filter
SPECT data were collected by scanning cylindrical phantom (inner height: 20 cm, inner diameter: 20.2 cm) loaded with 14.04 mCi of Tc-99m for system volume sensitivity and 20.41 mCi of Tc-99m for uniformity test without and with physical filter, respectively. The experimental setup is shown in the Figure 3. The radius of rotation of gamma camera head was 15 cm. The physical filter was mounted on the face of LEHR collimator. Uniformity images were reconstructed with FBP method by applying the Butterworth filter of order 5 and cut-off frequency 0.35 cycles/cm and Chang's attenuation correction method using 0.13 cm^{-1} linear attenuation coefficient was applied.

3. Results and discussion

3.1. Tc-99m spectra without and with physical filter
For analysis purposes, photopeak energy window was divided into three sub-windows, i.e., (i) 126-139 keV, (ii) 140 keV and (iii) 141-154 keV (Figure 5(a)). Figure 5(b) shows the average percentage reduction from sub-window (i) for 0.35 mm physical filter was 29.49% and for 0.40 mm physical filter was 40.81%. While in sub-window (ii), the average percentage reduction for 0.35 mm physical filter

was 38.76% and for 0.40 mm physical filter was 45.16%. For sub-window (iii), the average percentage reduction for 0.35 mm physical filter was 44.89% and for 0.40 mm physical filter was 50.41%. The higher percentage reduction in sub-window (ii) shows that Fe with 0.35 mm and 0.40 mm thicknesses are not suitable as a physical filter because more unscattered gamma photons were reduced. Improvement might have if the preferred thickness of the physical filter was used to remove more scattered gamma photons.

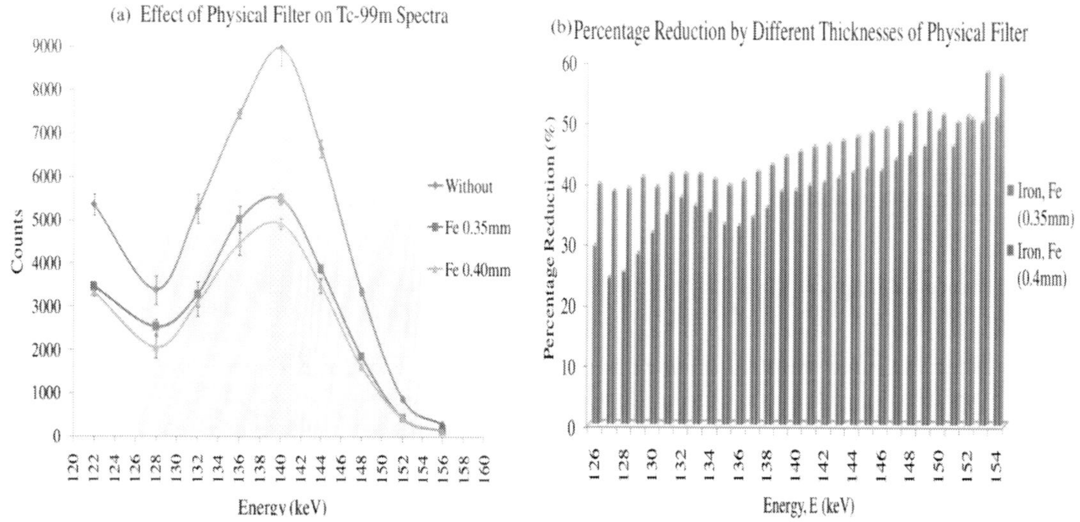

Figure 5. (a) The effect of Fe as a physical filter on Tc-99m spectra and (b) The percentage reduction by different thicknesses of physical filter.

3.2. *Spatial resolution without and with physical filter in air and scattering medium*

The full-width-half-maximum (FWHM) was measured for without and with physical filter. From Figure 6, the FWHM for 0.35 mm physical filter was improved as source-to-collimator distance increases from 1 to 3 cm compared to without physical filter. There was also an improvement in FWHM with 0.40 mm physical filter from 1 to 4 cm source-to-collimator distance. The FWHM for both 0.35 mm and 0.40 mm physical filter was poor as source-to-collimator distance increased from 5 cm and above.

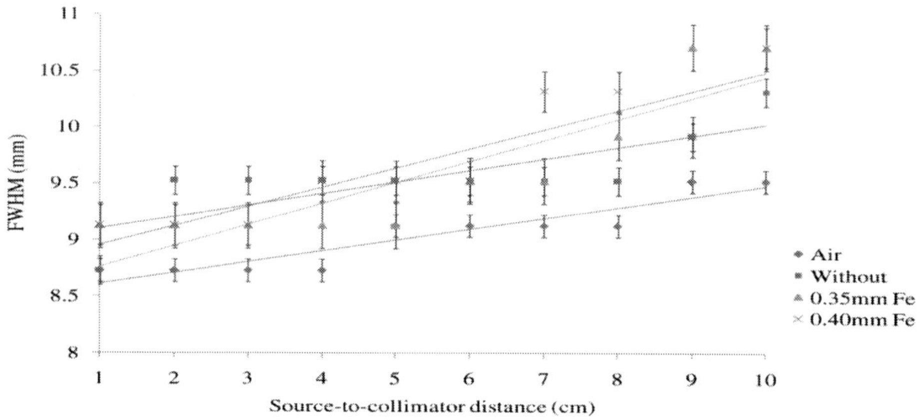

Figure 6. Spatial resolution without and with physical filter.

3.3. *System volume sensitivity and uniformity without and with physical filter*

The SPECT data were interpreted and analysed by using Pegasys software. The results show that the system volume sensitivity was 0.1677 cps/Bq/ml without applying physical filter. For the 0.35 mm and 0.40 mm physical filter, the system volume sensitivities were 0.1618 cps/Bq/ml (3.52%) and 0.1594 cps/Bq/ml (4.95%), respectively. There was reduction in the system volume sensitivity compared to without physical filter.

For uniformity, a straight line was drawn through the reconstructed image to obtain count profile. There is fluctuation in count profile for both thicknesses as well as without physical filtered data images and no improvement (Figure 7). This is probably due to unsuitable of chosen physical filter thicknesses.

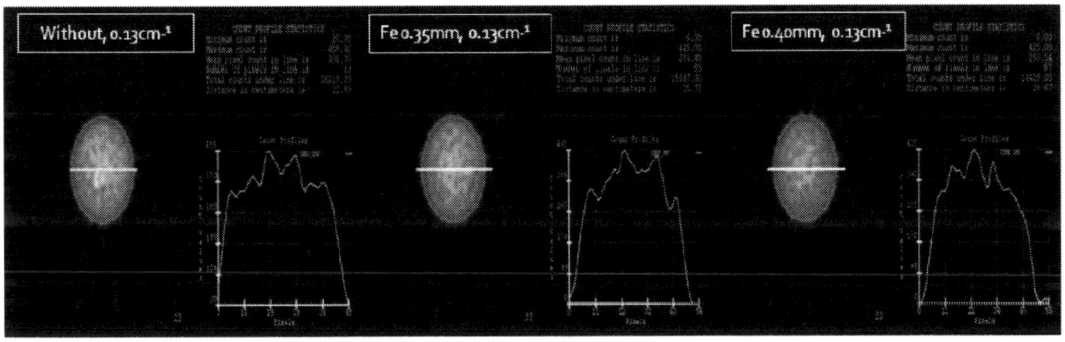

Figure 7. Count profiles for without and with Fe as a physical filter.

4. Conclusion

Results show the significant reduction in count rate with physical filter from photopeak region of Tc-99m spectra. At certain source-to-collimator distance, spatial resolution was improved. However, system volume sensitivity was reduced. For the uniformity, no improvement has been achieved. Only on the basis of spatial resolution results at 1 to 4 cm source-to-collimator distance, these thicknesses of physical filter may be tested further by scanning different planar/SPECT phantoms in Tc-99m imaging.

Acknowledgement

Authors acknowledge the support and generosity of School of Health Sciences, Universiti Sains Malaysia and Department of Nuclear Medicine, Radiotherapy and Oncology, Hospital Universiti Sains Malaysia for the facility, without which the present study could not have been completed.

References

[1] Ficke DC and Ter-Pogossian MM 1990 Hardening of Annihilation Radiation for Improved Coincidence Detection in PET *Conference Record IEEE Medical Imaging Conference* 1303-1304

[2] Guido Germano 2001 Technical aspects of myocardial SPECT imaging *J Nucl Med Tech* **29** 1499-1507

[3] Gustafson A, Arlig A, Jacobssen L, Lungberg M and Wikkelso C 2000 Dual energy window scatter correction and energy window setting in cerebral blood flow SPECT *Phys Med Biol 2000* **45** 3431-3440

[4] Hubbel, J.H. and Berger, M.J. May 10 1965 Photon Attenuation and Energy Transfer Coefficients. Tabulation and Discussion. United States Department of Commerce. *National Bureau of Standards Report 8681*

[5] Kojima A, Matsumoto M, Takahashi M. 1991 Experimental analysis of scattered photons in a 99mTc imaging with a gamma camera *Ann Nucl Med 1991* **5** 139-144

[6] Pillay, M., Shapiro, B., Cox, P.H. 1986 The Effect of an Alloy Filter on Gamma Camera Images *J Nucl Med* **12** 293-295

[7] Shah SI. 1996 Reduction of scattered gammas and attenuation correction in Tc-99m SPET imaging *Ph.D Thesis* (University of London, UK)

[8] Shah SI, Ahmad Zakaria and Tuhin Haque 2011 A New Technique for Reduction of

Scattered Gamma Photons in Tc-99m SPECT Imaging *Poster presentation in IIUM Research, Invention and Innovation Exhibition (IRIIE 2011)* Unpublished

[9] Spinks TJ and Shah SI 1993 Effect of lead filters on the performance of a neuro-PET tomograph operated without septa *IEEE Transactions on Nuclear Science* Aug 1993 **40** 1087-1091

[10] Zaidi H. 2006 *Quantitative analysis in nuclear medicine imaging* (Springer) p 368

Dosimetric characteristics of fabricated silica fibre for postal radiotherapy dose audits

M S Ahmad Fadzil[1], N N H Ramli[1], M A Jusoh[1], T Kadni[2], D A Bradley[3], N M Ung[4], H Suhairul[5] and N Mohd Noor[1]

[1] Department of Imaging, Faculty of Medicine and Health Sciences, Universiti Putra Malaysia, 43400 Serdang, Selangor, Malaysia
[2]Secondary Standard Dosimetry Laboratory Malaysia, Malaysian Nuclear Agency, Bangi, 43000 Kajang, Selangor, Malaysia
[3] Centre for Nuclear and Radiation Physics, Department of Physics, University of Surrey, Guildford, GU2 7XH, UK
[4] Clinical Oncology Unit, Faculty of Medicine, University of Malaya, 50603 Kuala Lumpur, Malaysia
[5] Department of Physics, Universiti Teknologi Malaysia, 81310, Skudai, Johor, Malaysia

E-mail: noramaliza@upm.edu.my

Abstract. Present investigation aims to establish the dosimetric characteristics of a novel fabricated flat fibre TLD system for postal radiotherapy dose audits. Various thermoluminescence (TL) properties have been investigated for five sizes of 6 mol% Ge-doped optical fibres. Key dosimetric characteristics including reproducibility, linearity, fading and energy dependence have been established. Irradiations were carried out using a linear accelerator (linac) and a Cobalt-60 machine. For doses from 0.5 Gy up to 10 Gy, Ge-doped flat fibres exhibit linearity between TL yield and dose, reproducible to better than 8% standard deviation (SD) following repeat measurements ($n = 3$). For photons generated at potentials from 1.25 MeV to 10 MV an energy-dependent response is noted, with a coefficient of variation (CV) of less than 40% over the range of energies investigated. For 6.0 mm length flat fibres 100 μm thick x 350 μm wide, the TL fading loss following 30 days of storage at room temperature was < 8%. The Ge-doped flat fibre system represents a viable basis for use in postal radiotherapy dose audits, corrections being made for the various factors influencing the TL yield.

1. Introduction

From 1969 until present time, the International Atomic Energy Agency (IAEA) in collaboration with the World Health Organization (WHO) have utilised a conventional phosphor-based transfer dosimeter TLD system as a dose audit tool. This forms an independent quality audit for the dose delivered by external beam radiotherapy treatment machines [1]. The limitations of such dosimeters include potential hygroscopic problems and thermal fading [2]. This has stimulated studies of various alternative materials for medical radiation TL dosimeters [3]. A study [4] has suggested that one of the credible TL materials is silica dioxide (SiO_2) optical fibre. This is due to its good spatial resolution; ~ 0.1 mm [3], water and corrosion resistance as well as having low residual signal and modest cost [5].

The main dosimetric characteristics of TLD of interest are linearity, reproducibility, energy dependence and fading. Previous study [5] has concluded that commercially available Ge-doped cylindrical cross-section silica optical fibres (~ 0.1 mm diameter) provide a good basis for medical radiation dosimetry at therapy levels. In present study use has been made of various sizes of fabricated flat fibres (of sub-mm cross-sectional dimension) doped with germanium (6 mol %), this doped medium exhibiting TL yield significantly greater than that of telecommunication fibres doped with Al, Nd, Yb, Er and Sm [3][6]. As such, there is the expectation that it can provide a sensitive system for therapy dose audit, approaching or perhaps exceeding that of TLD-100.

Prior to use for dose audit, there is a need to establish the characteristics of the fabricated dosimeter for accurate measurement and compliance to international codes of practice and the standard procedure used by the IAEA [7] and other agencies in postal dose audit programmes. This is the focus of present research.

2. Materials and methods

2.1. Sample fabrication

Various size of Ge-doped flat fibres were used in this study, the fibres comprising rectangular samples all of 6 ± 1 mm length, with thicknesses and widths of 60 x 180 µm, 85 x 270 µm, 100 x 350 µm, 165 x 620 µm and 200 x 750 µm. The fibres were produced using the modified chemical vapour deposition (MCVD) method, MCVD being carried out at the Multimedia University, with subsequent fibres production using the pulling-tower facilities of the Photonics Research Group, Department of Electrical Engineering, University of Malaya. Forming of the flat fibres was performed by applying a vacuum to the doped hollow silica preform during the fiber drawing process [8].

2.2. Preparation of Ge-doped flat fibres

The fibres were cleaned using methanol before cutting the fibres into their respective sizes using a S90R diamond cutter (Thorlabs, USA). The mass of each optical fibre dosimeter was obtained using an AB204-S electronic balance (Mettler Toledo, Canada) as shown in Table 1. The fibre mass was used to normalize the thermoluminescence (TL) signal, obtain TL yield per unit mass. In order to minimize surface abrasion of the fibre and deposition of dust or finger oil, vacuum tweezers (Dymax 5, Surrey, UK) were used to handle the optical fibres.

Table 1. Average mass of fibre.

Size of fibre (cladding + core)	Average mass of fibre (g)
60 x 180 µm	0.0035
85 x 270 µm	0.0044
100 x 350 µm	0.0053
165 x 620 µm	0.0097
200 x 750 µm	0.0109

2.3. Annealing

To erase any pre-irradiation TL signal (due for instance to tribological exposure) the optical fibres were annealed using a furnace (Carbolite, Derbyshire, UK). The fibres were placed in a brass holder and annealing was performed at a temperature of 400 °C for 1 hour, followed by a period of 8 hours of slow cooling to avoid thermal stress. After cooling, the fibres were placed inside a black light-tight container in order to minimize exposure to ambient light.

2.4. Sample irradiation

The irradiations were carried out using a Co-60 source located at the Secondary Standard Dosimetry Laboratory of the Nuclear Malaysia Agency and also using a University of Malaya Medical Centre Varian electron linac operated at 6 MV and 10 MV to deliver photon irradiations at commensurate energies. Five gelatine capsules were each loaded with four flat fibre dosimeters of 6.0 ± 1.0 mm length, each capsule containing a particular cross-sectional set of fibres with thicknesses and widths as follows: 60 x 180 μm, 85 x 270 μm, 100 x 350 μm, 165 x 620 μm and 200 x 750 μm.

2.4.1. Linearity and energy dependence. The optical fibres were irradiated to doses in the range 0.5 Gy up to 10 Gy at a dose rate of 600 monitor unit (MU) per minute (for present conditions, 6 Gy per minute). During gamma irradiation using the Co-60 facility, the five different sizes of fibre were held in a polythene zipper bag placed at the surface of a Perspex (Polymethyl methacrylate, PMMA) phantom of dimensions (30 x 15 x 30) cm^3 with a source-to-surface distance (SSD) of 100 cm and a field size of 10 x 10 cm^2. For the 6 MV and 10 MV linac irradiations, the capsules were similarly held in a zipped storage bag and placed on a solid-water™ phantom, this time of dimensions (20 x 20 x 10) cm^3, to then be sandwiched with 1.5 cm bolus. The focus to surface distance (FSD) was 100 cm and a 10 x 10 cm^2 field size was used. Mean measured doses for each fibre group have been used to obtain a plot against delivered doses obtained from ion chamber measurements.

2.4.2. Reproducibility. For this, a total of four optical fibres were irradiated at dose rates of 0.0427 Gy min^{-1} and 600 cGy min^{-1} for Co-60 and the linac irradiations (6 MV and 10 MV photons) respectively, with doses ranging from 1 Gy to 2 Gy. Two repeat irradiations were made in order to assess the reproducibility of the flat fibres at the selected energy and dose.

2.4.3 Fading. Evaluation was made of the stability of TL signal over the period 1 to 30 days post irradiation. Irradiated flat fibres, dosed to 2 Gy, were read-out upon completion of the 1st, 6th, 16th and 22nd day post-irradiation.

2.5. Readout
The samples were read-out using a Harshaw 3500 TLD reader with a nitrogen gas atmosphere of 0.5 bars, used to suppress spurious light signals from triboluminesence and to reduce surface oxidation [9]. The following time-temperature profile was used during readout: preheat temperature of 180°C for 5 seconds; readout temperature of 400°C for 6 seconds and heating cycle rate of 35°C per second. An internal annealing temperature of 400°C was used for 10 seconds to remove any residual (deep traps) signal in the fibres.

3. Results and discussion

3.1. TL response

3.1.1. Linearity. The TL response of the fibres was found to be linear over the investigated dose range, as shown in Figures 1 to 3. Based on the coefficient of determination (R^2) of each flat fibre, the fibres of dimension 85 x 270 μm, 60 x 180 μm and 100 x 350 μm provided the best agreement between TL yield per unit mass and delivered dose for the Co-60 and 6 MV and 10 MV linac photons respectively. In line with expectation, based on sample mass, the least gradient corresponds to 60 x 180 μm for all photon energies delivered. For the thickest sample it is apparent that the additional attenuation of TL light transported through the medium suppresses any additional benefit of increased sample mass. The suppression becomes increasingly apparent with dose. This behaviour is a consistent feature that is observed in Figures 1 to 4.

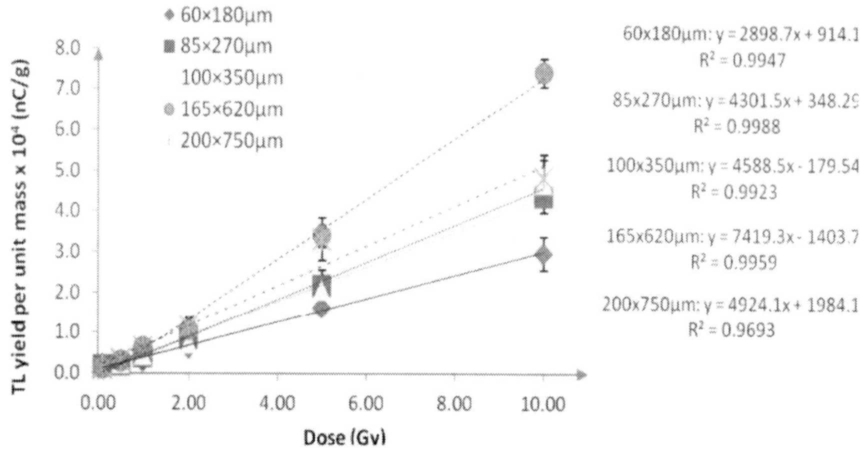

Figure 1. Linearity of the Ge-doped flat fibres for the 1.25 MeV mean energy of the Co-60 gamma beam irradiation.

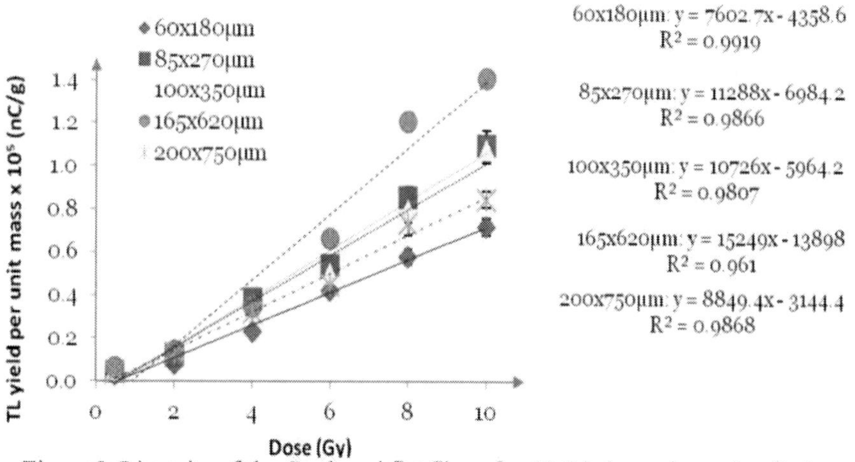

Figure 2. Linearity of the Ge-doped flat fibres for 6 MV photon beam irradiation.

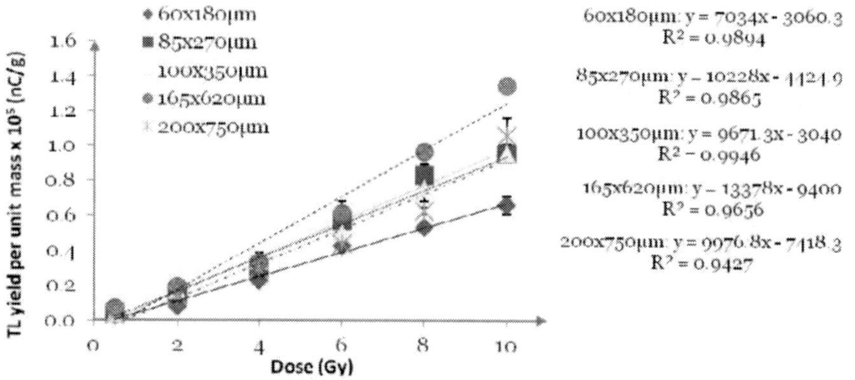

Figure 3. Linearity of the Ge-doped flat fibres for 10 MV photon beam irradiation.

3.1.2. Energy dependence. As shown in Figure 4, the Ge-doped flat fibres employed in this study exhibit increasing response at 6 MV compared to that obtained from Co-60 irradiation. However, the TL response obtained for the 10 MV photon beam is lower than that obtained at 6 MV, being in accord with [9] who observed the 9 μm commercial Ge-doped fibres response is decreased for 10 MV compared to 6 MV irradiations due to the mass energy absorption coefficient at 6 MV being higher than that for 10 MV. Overall, the results in this study reveal small energy dependence, with a coefficient of variation (CV) of less than 40% over the range of energies investigated.

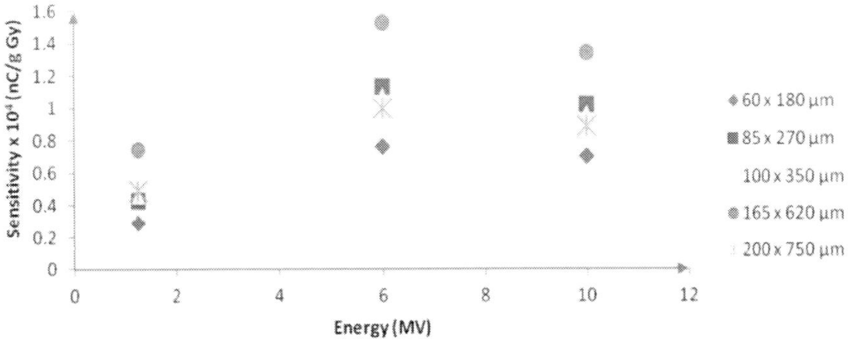

Figure 4. Energy responses of Ge-doped flat fibres following 1.25 MeV, 6 MV and 10 MV photon irradiations at a dose of 2 Gy.

3.1.3. Reproducibility. In this regard, the Ge-doped flat fibres were irradiated at a dose rate of 600 cGy min^{-1} for both 6 MV and 10 MV photons. Table 2 shows the mean TL yield per unit mass of each size of flat fibre for the corresponding doses and number of fibre samples read-out. The dose delivered was 1 Gy for Co-60 irradiations and 2 Gy for the linac irradiations. The standard deviation of these flat fibres in percentage terms was generally within 1.25 to 7.56%.

Table 2. Dose reproducibility of 1.25 MeV, 6 MV and 10 MV.

Size of flat fiber (μm)	Energy (source)	Dose (Gy)	No of fiber read	TL yield per unit mass (nC/g)		
				Mean	SD	SD in %
60 x 180	1.25 MeV	1	4	3857.58	106.39	2.76
	6 MV	2	4	5159.55	319.33	6.19
	10 MV	2	4	5080.00	301.26	5.93
85 x 270	1.25 MeV	1	4	3398.88	42.48	1.25
	6 MV	2	4	4783.85	212.98	4.45
	10 MV	2	4	6353.85	285.65	4.50
100 x 350	1.25 MeV	1	4	3897.58	151.74	3.89
	6 MV	2	4	7116.76	439.66	6.18
	10 MV	2	4	8294.78	288.18	3.47
165 x 620	1.25 MeV	1	4	18481.25	500.66	2.71
	6 MV	2	4	25427.08	1110.40	4.37
	10 MV	2	4	36333.33	919.13	2.53
200 x 750	1.25 MeV	1	4	7329.95	224.90	3.07
	6 MV	2	4	682.91	51.60	7.56
	10 MV	2	4	897.47	35.47	3.95

3.1.4. Fading. The loss of TL signal of irradiated flat fibres at 1st, 6th, 16th and 22th day post-irradiation is shown in Figure 5 and Figure 6 for 6 MV and 10 MV respectively. The line linking points are provided as a guide to the eye but not fit to each point. After 22 days, the largest signal loss was 51.11 %, suffered by the 200 x 750 μm fibre irradiated using 6 MV photons while the 165 x 620 μm fibre which been irradiated with 10 MV photon showed a signal loss of 44%. Conversely, the 100 x 350 μm fibre showed the least TL signal loss, at 6.55%, a matter worthy of continuing research in an effort to better understand the source of variation.

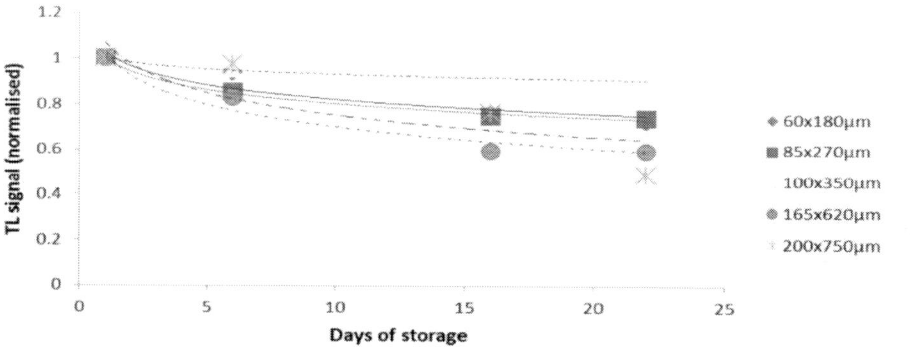

Figure 5. Decay fit to 4 measured fading values for 6 MV. Each data point corresponds to the average reading of 4 flat fibres.

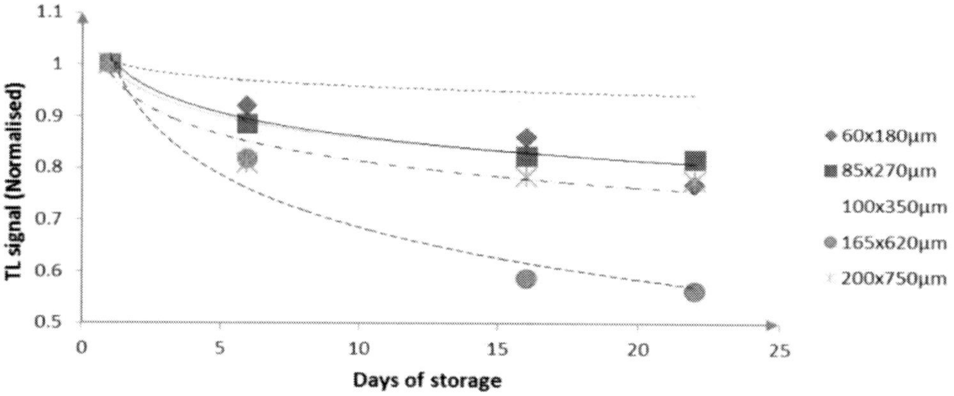

Figure 6. Decay fit to 4 measured fading values for 10 MV. Each data point corresponds to the average reading of 4 flat fibres.

4. Conclusion

The dosimetric characteristics of novel fabricated Ge-doped flat fibres have been established. Investigation reveals the 85 x 270 μm flat fibre to have the best performance, offering a viable system for postal radiotherapy dose audits, with corrections required for energy dependence, reproducibility and fading, suggesting the correction methods based on the Ge-doped audit system developed by Noor et al [9] need to be employed in order to calculate the absolute absorbed dose from Ge-doped optical fibres measurement for lower and higher energy photon.

Acknowledgments

This research is supported by PUTRA Grant no: 9404200 from Universiti Putra Malaysia, Malaysia Fundamental Research Grant Scheme (FRGS) no: 02021313l7FR and Universiti Malaya High Impact

Research Grants (UM-HIR) no UM.C/625/1/HIR/33 and A000007-50001 from the Ministry of Education Malaysia. Use of irradiation facilities at the Malaysian Nuclear Agency is acknowledged.

References

[1] Izewska J and Andreo P 2000 The IAEA/WHO TLD postal programme for radiotherapy hospitals. *Radiotherapy and Oncology : Journal of the European Society for Therapeutic Radiology and Oncology*, **54**(1), 65–72.

[2] Rah J R, Kim S, Cheong K., Lee J, Chung J, Shin D and Suh Tae 2009 Feasibility study of radiophotoluminescent glass rod dosimeter postal dose intercomparison for high energy photon beam. *Applied Radiation and Isotopes* **67**, 324-328.

[3] Hashim S, Bradley D A, Peng N, Ramli A T and Wagiran H 2010 The thermoluminescence response of oxygen-doped optical fibres subjected to photon and electron irradiations. *Nuclear Instruments and Methods in Physics Research Section A: Accelerators, Spectrometers, Detectors and Associated Equipment*, **619**(1-3), 291–294.

[4] Bradley D A, Hugtenburg R P, Nisbet A, Abdul Rahman A T, Issa F, Mohd Noor N and Alalawi A. 2012 Review of doped silica glass optical fibre: Their TL properties and potential applications in radiation therapy dosimetry. *Applied Radiation and Isotopes*, **71**, 2–11.

[5] Abdulla Y A, Amin Y M, and Bradley D A 2001 The thermoluminescence response of Ge-doped optical fibre subjected to photon irradiation. *Radiation Physics and Chemistry*, **61**(3-6), 409–410.

[6] Noor N M, Hussein M, Bradley D A and Nisbet A 2011 Investigation of the use of Ge-doped optical fibre for in vitro IMRT prostate dosimetry. *Nuclear Instruments and Methods in Physics Research Section A: Accelerators, Spectrometers, Detectors and Associated Equipment*, **652**(1), 819–823.

[7] P. Andreo, D. T. Burns, K. Hohlfeld, M. S. Huq, T. Kanai, F. Laitano, V. G. Smyth and S. Vynckier 2006 *Absorbed dose determination in external beam radiotherapy: An international code of practice for dosimetry based on standards of absorbed dose to water* Vienna, IAEA **12**, 13-20.

[8] Dambul K D, Tamchek N, Sandoghchi S R, Abu Hassan M R, Tee D C and Mahamd Adikan F R 2011 Fabrication and characterization of flat fibers. *2011 ^{2}nd International Conference on Photonics (ICP), Kota Kinabalu,* 1–4.

[9] Noor N M, Hussein M, Kadni T, Bradley D A and Nisbet A 2014 Characterization of Ge-doped optical fibres for MV radiotherapy dosimetry. *Radiation Physics and Chemistry*, **98**, 33–41.

Raman study of lower toxicity polymer gel for radiotherapy dosimetry

M Z Adenan[1,4], M Ahmad[2], N Mohd Noor[3], N Deyhimihaghighi[1] and E Saion[1]

[1] Department of Physics, Faculty of Science, Universiti Putra Malaysia, 43400 UPM Serdang, Selangor, Malaysia
[2] Department of Chemistry, Faculty of Science, Universiti Putra Malaysia, 43400 UPM Serdang, Selangor, Malaysia
[3] Department of Imaging, Faculty of Medicine and Health Science, Universiti Putra Malaysia, 43400 UPM Serdang Selangor, Malaysia
[4] Department of Medical Imaging, Universiti Teknologi MARA Puncak Alam Campus, Bandar Puncak Alam 42300 Kuala Selangor, Selangor, Malaysia

E-mail: ibnu_soleh83@yahoo.com

Abstract. N-isopropyl acrylamide (NIPAM) monomer and *N, N'* – methylene-bis-acrylamide (BIS) crosslinker were used to synthesize polymer gel dosimeters for a reason that the monomer is lower toxicity which gives a significant advantage over the other polymer gel compositions. The gels were irradiated with Co-60 gamma rays at doses up to 21 Gy and the irradiated NIPAM polymer gels were used to investigate the dose response characteristics based on Raman spectroscopy analysis on the formation of the polymer gels and the consumptions of NIPAM and BIS co-monomers. From the findings, the polymerization was referred to an increment in Raman intensity at 815 cm^{-1}, assigned for C-C stretching mode of NIPAM polymer gel, as the dose increased. The consumptions of the co-monomers were referred to a decrement in Raman intensities at 1025 cm^{-1} and 2353 cm^{-1} for C=C stretching modes of NIPAM and BIS respectively as the dose increased. The increment and decrement in Raman intensities of polymer and co-monomers respectively with increase of dose indicate that there is occurrence of polymerization of NIPAM polymer gels which could be applied in 3D dose distributions for radiotherapy treatment planning. The correlation factor k_{BIS} is greater than k_{NIPAM}, showing that the reaction of BIS crosslinker is more efficient than NIPAM monomer to generate 37% of the NIPAM polymer gel.

1. Introduction

Research in polymer gel dosimeter has been undergone as a subject of interest from the past 20 years, iniated by Maryanski [1], in which there is a kind of demand to develop the better dosimeter with high spatial resolution and three-dimensional capability for the application in radiotherapy treatment plan. Stated by Mc Auley et. al. [2-4], the polymer gel has been used by medical physicists to verify spatial three-dimensional dose distribution treatment plans before starting the actual radiotherapy procedure which focused to only irradiate the target tissue or organ while keeping away the surrounding tissue from radiation. The conventional dosimeter such as thermoluminiscent dosimeter can quantify radiation in one or two direction only, thus limit the function in the complex dose application in the radiotherapy procedure. Polymer gel dosimeter is fabricated from radiation sensitive chemicals known as monomers which are upon irradiation, the monomers will polymerize as a function of absorbed radiation dose [2], thus the polymerized monomers are held firmly in the gelatine. The previous polymer gels has been prepared under hypoxic conditions [5], until study made by Fong et. al. [6, 7], then polymer gel can be synthesized under normal atmospheric environment by applying certain amount of anti-oxidant agent due to its sensitivity to oxygen that may disturb the polymerized reaction. Study done by De Deene [8] named this new type of polymer gel as normoxic instead of anoxic for the old type.

One of major limitations in establishing polymer gel dosimeter in routine clinical practice is the problem of toxicity in monomers [9, 10]. Therefore some studies put their concern to discover new monomers with lower toxicity as well as easy in fabrication under normal atmospheric environment [10, 11]. Application of N-isopropyl acrylamide (NIPAM) monomer in polymer gel composition together with *N, N'* –methylene-bis-acrylamide (BIS) as crosslinker was first introduced by Senden [10], which is relatively less toxicity from the monomer applied in previous literature such as normoxic polyacrylamide gels (nPAG) [12]. Study done by Mesbahi [13] which applied the clinical magnetic resonance imaging (MRI) scanner as data acquisition methods, proven that the R_2-dose response of NIPAM polymer gel was comparable to normoxic polyacrylamide gel (PAGAT) established by Senden [10] which has high dose sensitivity and less dependent in dose rate and irradiation temperature. In fact, Mesbahi [13] observed that, NIPAM polymer gel was better than PAGAT gel in linearity, since the dose-response (R_2) of NIPAM was significantly linear within larger dose range.

In terms of dose response analysis, once the polymer gel dosimeter has been irradiated, it can be evaluated by using certain modalities including MRI [3, 13, 14], X-ray computed tomography (X-ray CT) [7, 9, 15], optical CT [7, 10, 16], ultrasound[17] and Raman spectroscopy [5, 18, 19]. Thus, the relative dose response information can be extracted for further analysis. In this study, Raman spectroscopy has been chosen to evaluate and probe the changes in molecular structure of the polymer gels following irradiation. NIPAM polymer gel which has certain advantages is a main concern to be further investigated by using Raman spectroscopy in monitoring the characteristic set of covalent bond on each molecule within the gel dosimeter. In particular, several batches of NIPAM polymer gels with different sets of monomer concentration are to be characterized by Raman spectroscopy to evaluate the changes in molecular structure of the gel sample upon irradiation process. This study may focus with the information of Raman spectra that may reflect the activity of polymer formation and monomer consumption [19].

2. Materials and methods

2.1. NIPAM polymer gel preparation for Raman spectroscopy analysis

Polymer gel dosimeters were manufactured under normal atmospheric conditions in the laboratory without using of a glove box based on Fong et. al. [5, 6] and basically following the method experimented by Senden et. al. [10, 20]. Firstly, gelatine (6 wt%) (bovine skin, Type B, Sigma Chemical Co) was added to the deionized water and stirred for five minutes at room temperature. Then, gelatine solution was heated up to 45°C and continuously stirred by magnetic stirrer until a clear solution and transparent was obtained. With heating was off and continuous stirring was pursued, the monomer and co-monomer, BIS (2-4 wt%) (Sigma Chemical Co) and NIPAM (2-4 wt%) (Sigma Aldrich) were poured into the gelatine solution and approximately dissolved within 15 minutes. Then the anti-oxidant, tetrakis hydroxymethyl phosphonium chloride (THPC) (5mM) was added to solution and continuous stirred at about 2 minutes. The final product was filled into small ampoule tubes (2ml), which was sealed with parafilm tape and covered by aluminium wrapped container to minimize oxidation and photo-polymerization effect. Upon completion, the prepared NIPAM polymer gel dosimeters were stored in a refrigerator at low temperature approximately 4°C to solidify and protected from UV light. Once the polymer gel dosimeters were ready, the next day (within 24 hours), gels were removed from refrigerator and equilibrate at room temperature before starting irradiation process.

2.2. NIPAM polymer gel irradiation

Because the gel samples were small, so the time to equilibrate the samples with room temperature took less than 10 minutes. Then, the polymer gel dosimeters were irradiated with gamma rays produced by Cobalt-60 (Co^{60}) element from Gamma Cell Instrument, model 220 Excel Irradiator (MDS Nordion, Ottawa, ON, Canada) (Faculty of Nuclear Science, National University of Malaysia). The amount of radiation dose given was set from a control panel of the irradiation chamber. One gel sample from each batch of dosimeters was left unirradiated and just put inside the refrigerator, as a control sample and the rest were irradiated batch by batch simultaneously according to the amount of radiation dose which already calculated ranging from 3 Gy to 21 Gy with a single dose at a time. During irradiation, each sample was placed in a beaker filled with a distilled water to create tissue equivalent medium.

2.3. Raman spectroscopy of NIPAM polymer gel

All gel samples were characterized by using Raman spectrometer (RSI 2001, Raman system, INC) equipped with 532 nm solid-state diode green laser and a thermoelectrically cooled CCD array after seven days of irradiation in order to allow complete polymerization of gel samples [8]. Data manipulation and analysis were established by using Grams/32 software, version 6. All spectra were corrected for base line, smoothing and Fourier Transform (FT). A constant correction factor of 80% of the degree of smoothing parameter was used throughout the data collection.

3. Results and discussion

3.1. Raman peaks identification

Raman spectra that obtained from Raman spectrometer was actually the vibrational Raman effect in which it was directly described the polymer formation and co-monomers consumption after irradiation. From the Raman analysis, the relative intensity which represents the energy difference of scattered radiation can be acquired and displayed on the chart as a function of difference in wavelength. The well define peak on the chart may represent characteristic molecular bond in the gel samples which has certain vibrational energy. The characteristic frequencies of every single functional group were characterized by referring the literature reports documented by Jirasek et. al. [18, 19]. The polymerization was followed from the increment in Raman intensity at 815 cm^{-1}, assigned for C-C stretching mode of NIPAM polymer gel, as the dose increased. Meanwhile, the consumptions of the co-monomers were referred to a decrement in Raman intensities at 1025 cm^{-1} and 2353 cm^{-1} assigned for C=C stretching modes of NIPAM and BIS respectively, as the dose increased. These three chosen peaks were obtained from Raman spectra which were illustrated in Figure 1.

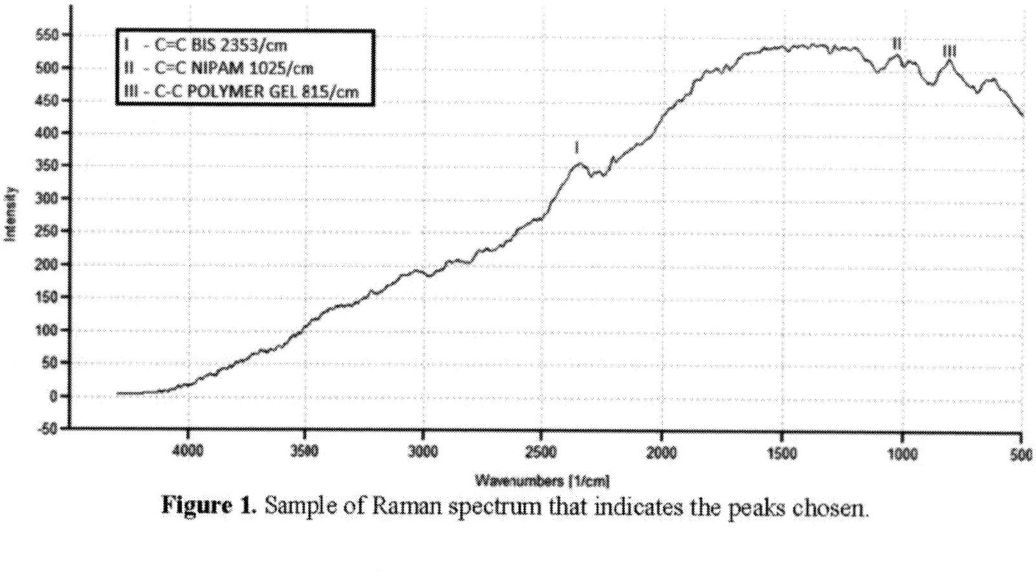

Figure 1. Sample of Raman spectrum that indicates the peaks chosen.

3.2. Polymer formation of NIPAM polymer gel

The Raman intensities, y, as a function of dose D for NIPAM and BIS were fit [19] and displayed as monoexponential function, following the equation:

$$y = y_0 + A(1 - e^{-D/D_0})$$

Where, D_0 is the sensitivity parameter, y_0 is the Raman intensity at zero dose and $A = y_{max} - y_0$ is the maximum differential dose response. y_{max} describes the maximum intensity over the dose range up to 21 Gy. Figure 2(a) and 2(b) depicts the positive monoexponential plots of the intensity change as a function of absorbed dose at initial concentrations of NIPAM monomer and different BIS crosslinker.

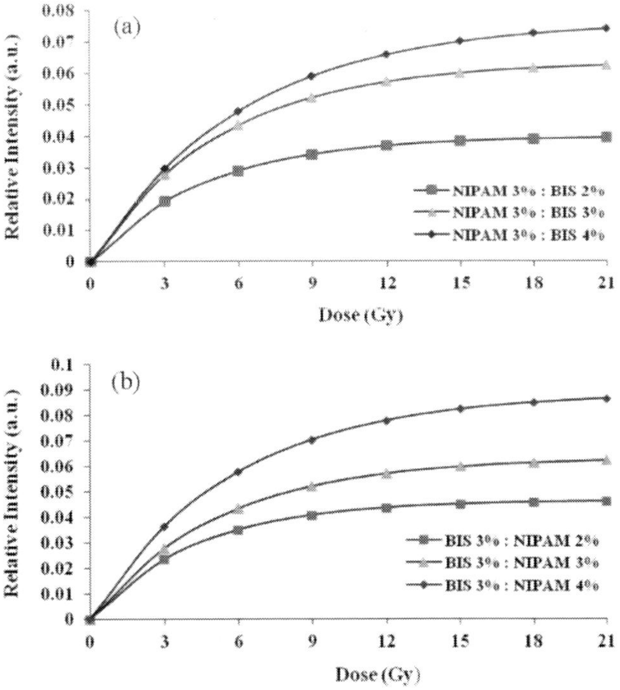

Figure 2. Data displayed a monoexponential function of relative Raman intensity versus absorbed dose for polymerization of NIPAM polymer gels of C-C bond at 815 cm^{-1} for (a) 3% NIPAM at various BIS from 2 to 4% and for (b) 3% BIS at various NIPAM from 2 to 4%.

Note that, as the radiation dose increases, the Raman intensity also increases, which explains the occurrence of polymerization process and the amount of carbon single bond increases in the formation of NIPAM polymer gel. All the carbon covalent bond tends to break into carbon single bond, as the radiation dose increases. The similar situation occurs, as the concentration of NIPAM monomer and BIS crosslinker increases. The dose sensitivity parameter D_0 can be determined from the reciprocal of the gradient of a linear plot $\ln\left[1 - \frac{\Delta y}{A}\right]$ versus dose D. This parameter can be described as the amount of dose required to generate 37% of the polymer formation. Thus, in Figure 3(a) and 3(b) indicates the linear correlations that may produce the gradients k_{NIPAM} and k_{BIS} respectively. It is shown that, k_{BIS} is greater than k_{NIPAM}, which proofs that the reaction of BIS is more effective than NIPAM in producing 37% of NIPAM polymer gel.

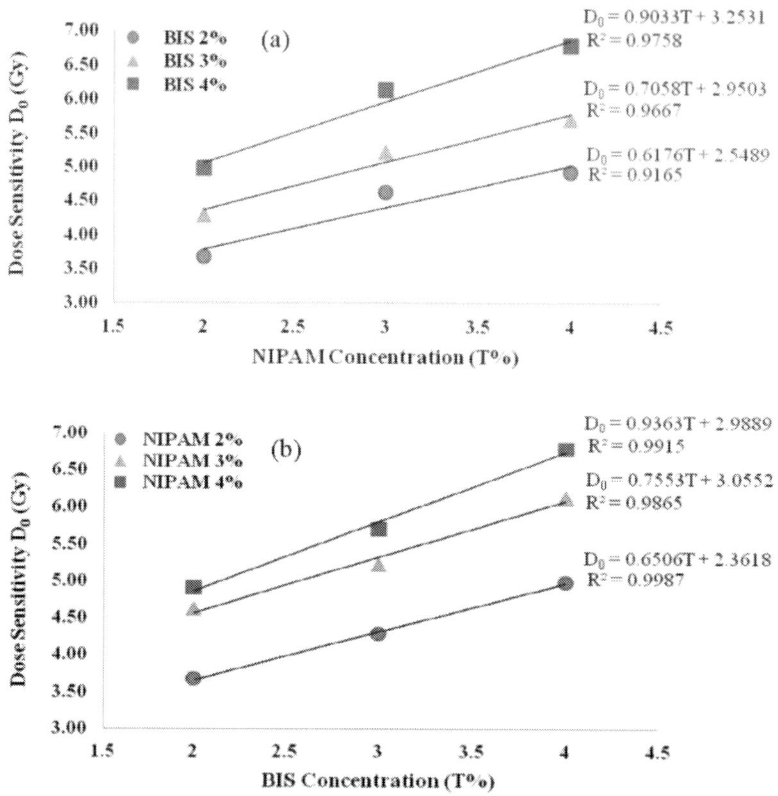

Figure 3. The polymerization process depicts the linear correlation between (a) D_0 vs. % NIPAM at various BIS from 2 to 4% and (b) D_0 vs. % BIS at various NIPAM from 2 to 4%.

3.3. Consumption of NIPAM and BIS

The Raman intensities, y, as a function of dose D for NIPAM and BIS were fit [4] and displayed as monoexponential function, following the equation:

$$y = y_0 - A(1 - e^{-D/D_0})$$

Where, D_0 is the sensitivity parameter, y_0 is the Raman intensity at zero dose and $A = y_{max} - y_0$ is the maximum differential dose response. y_{max} describes the maximum intensity over the dose range up to 21 Gy. By applying the equation above, it is seen in Figure 4(a) and 4(b) that, there is decreasing in monoexponential function of the intensity as a function of absorbed dose at the initial concentrations of NIPAM and BIS respectively from 0 Gy up to 21 Gy.

It is shown that, as the radiation dose increases, the Raman intensity decreases, which explains the NIPAM and BIS consumption increase and the amount of carbon covalent bond decreases in the formation of NIPAM polymer gel. All the carbon covalent bond tends to be broken into carbon single bond following irradiation process. The same pattern occurs, as the concentration of NIPAM monomer and BIS crosslinker increases. Figure 5(a) and 5(b) indicates the linear correlations that may generate the gradients k_{NIPAM} and k_{BIS} respectively.

Note that, k_{BIS} is greater than k_{NIPAM}, which explains that the consumption of BIS is at greater rate than NIPAM in developing 37% of NIPAM polymer gel. Therefore, generally, polymer gel that composed higher concentration of BIS is considerably better than the rest compositions towards the polymerization process.

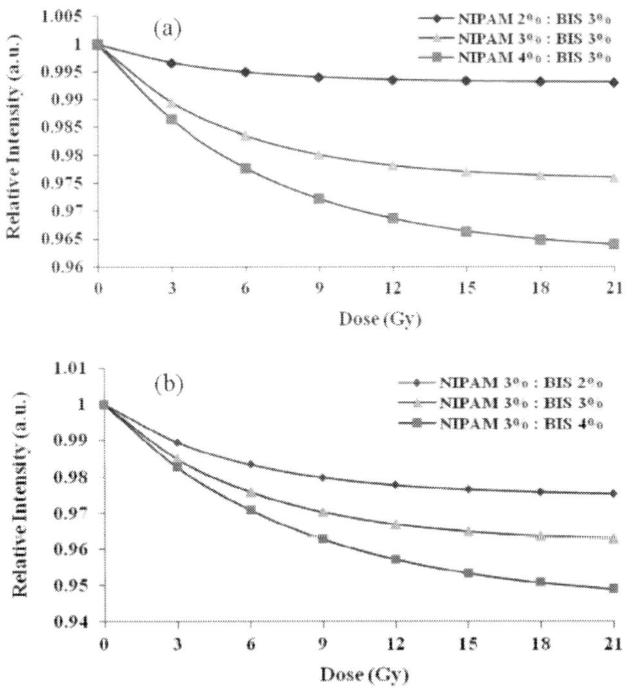

Figure 4. Data displayed a monoexponential function of relative Raman intensity versus absorbed dose for (a) 3% BIS at various NIPAM from 2 to 4% for consumption of C=C bond of NIPAM at 1025 cm^{-1} and for (b) 3% NIPAM at various BIS from 2 to 4% for consumption of C=C of BIS at 2353 cm^{-1}.

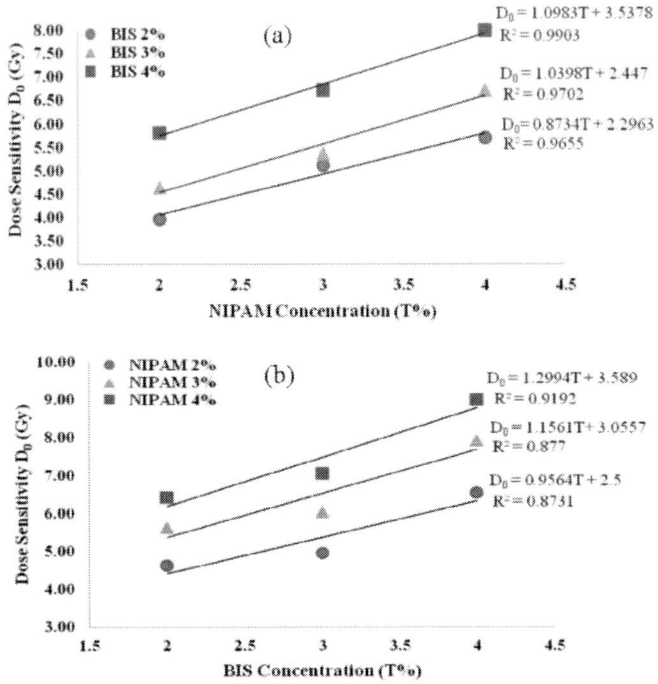

Figure 5. The co-monomer consumption process depicts the linear correlation between (a) D_0 vs. % NIPAM at various BIS from 2 to 4% and (b) D_0 vs. % BIS at various NIPAM from 2 to 4%.

4. Conclusion

The NIPAM polymer gels which composed mainly with NIPAM monomer and BIS crosslinker have been synthesized to study their characterization in terms of polymer formation and monomer consumption by using Raman spectroscopy. The results have shown that, there are changes in the Raman intensities as a function of absorbed dose, in which it proofs that the radiation may induce the gel samples to be polymerized. The increment and decrement in Raman intensities as the radiation dose increases indicates that there is occurrence of polymerization process and the co-monomers consumption process towards the formation of NIPAM polymer gel. The rate of polymerization increases as the amount of carbon single bonds increases and the rate of monomers consumption increases as the amount of carbon covalent bonds decreases after irradiation. In terms of dose sensitivity, it increases with the increment of NIPAM and BIS concentrations. In addition, the correlation factor, k_{BIS} is always greater than k_{NIPAM} whether in polymer formation and co-monomer consumption indicating BIS crosslinker is more efficient than NIPAM monomer toward the production of 37% of NIPAM polymer gel.

References

[1] Maryanski, M.J., et al., *NMR relaxation enhancement in gels polymerized and cross-linked by ionizing radiation: a new approach to 3D dosimetry by MRI*. Magnetic resonance imaging, 1993. 11(2): p. 253-258.

[2] McAuley, K.B. *Fundamentals of polymer gel dosimeters*. in *Journal of Physics: Conference Series*. 2006: IOP Publishing.

[3] Baldock, C., et al., *Polymer gel dosimetry*. Physics in medicine and biology, 2010. 55(5): p. R1.

[4] Rozlan, A.A., M.S. Jaafar, and A.A. Rahman, *Development and Characterization of Normoxic Polyhydroxyethylacrylate (PHEA) Gel Dosimeter using Raman Spectroscopy*.

[5] Rozlan, A.A., M.S. Jaafar, and A.A. Rahman. *A Study of Normoxic Polymer Gel Dosimeter Using Raman Spectroscopy Analysis*. in *Proceedings of International Conference on Biomedical Engineering and Technology (ICBET 2011)*. 2011.

[6] Fong, P.M., et al., *Polymer gels for magnetic resonance imaging of radiation dose distributions at normal room atmosphere*. Physics in medicine and biology, 2001. 46(12): p. 3105.

[7] Koeva, V., et al., *Preliminary investigation of the NMR, optical and x-ray CT dose–response of polymer gel dosimeters incorporating cosolvents to improve dose sensitivity*. Physics in medicine and biology, 2009. 54(9): p. 2779.

[8] De Deene, Y., et al., *A basic study of some normoxic polymer gel dosimeters*. Physics in medicine and biology, 2002a. 47(19): p. 3441.

[9] Ghavami, S.-M., et al., *Normoxic polymer gel dosimetry using less toxic monomer of N-isopropyl acrylamide and X-ray computed tomography for radiation therapy applications*. Reports of Practical Oncology & Radiotherapy, 2010. 15(6): p. 172-175.

[10] Senden, R., et al., *Polymer gel dosimeters with reduced toxicity: a preliminary investigation of the NMR and optical dose–response using different monomers*. Physics in medicine and biology, 2006. 51(14): p. 3301.

[11] Pappas, E., et al., *Narrow stereotactic beam profile measurements using N-vinylpyrrolidone based polymer gels and magnetic resonance imaging*. Physics in medicine and biology, 2001. 46(3): p. 783.

[12] De Deene, Y., et al., *The fundamental radiation properties of normoxic polymer gel dosimeters: a comparison between a methacrylic acid based gel and acrylamide based gels*. Physics in medicine and biology, 2006. 51(3): p. 653.

[13] Mesbahi, A., V. Jafarzadeh, and N. Gharehaghaji, *Optical and NMR dose response of N-isopropylacrylamide normoxic polymer gel for radiation therapy dosimetry*. Reports of Practical Oncology & Radiotherapy, 2012. 17(3): p. 146-150.

[14] Venning, A., et al., *Investigation of the PAGAT polymer gel dosimeter using magnetic resonance imaging*. Physics in medicine and biology, 2005. 50(16): p. 3875.

[15] Jirasek, A., M. Hilts, and K. McAuley, *Polymer gel dosimeters with enhanced sensitivity for use in x-ray CT polymer gel dosimetry*. Physics in medicine and biology, 2010. 55(18): p. 5269.

[16] Wuu, C.-S. and Y. Xu, *Three-dimensional dose verification for intensity modulated radiation therapy using optical CT based polymer gel dosimetry*. Medical physics, 2006. 33: p. 1412.
[17] Mather, M.L., A.K. Whittaker, and C. Baldock, *Ultrasound evaluation of polymer gel dosimeters*. Physics in medicine and biology, 2002. 47(9): p. 1449.
[18] Baldock, C., et al., *Fourier transform Raman spectroscopy of polyacrylamide gels (PAGs) for radiation dosimetry*. Physics in medicine and biology, 1998. 43(12): p. 3617.
[19] Jirasek, A., et al., *Characterization of monomer/crosslinker consumption and polymer formation observed in FT-Raman spectra of irradiated polyacrylamide gels*. Physics in medicine and biology, 2001. 46(1): p. 151.
[20] Hsieh, B., et al., *A study on dose response of NIPAM-based dosimeter used in radiotherapy*. Journal of Radioanalytical and Nuclear Chemistry, 2011. 290(1). p. 141-148.

9th National Seminar on Medical Physics (NSMP2014) IOP Publishing
Journal of Physics: Conference Series **546** (2014) 012012 doi:10.1088/1742-6596/546/1/012012

Characterisation of the thermoluminescence (TL) properties of tailor-made Ge-doped silica glass fibre for applications in medical radiation therapy dosimetry

N A Zahaimi[1], H Zin[2], G A Mahdiraji[3], A L Abdul Rahman[4], D A Bradley[5,6] and A T Abdul Rahman[1,5,7]

[1]Faculty of Applied Sciences, Universiti Teknologi MARA, 40450 Shah Alam, Malaysia
[2]Advanced Medical and Dental Institute, Universiti Sains Malaysia, Bertam, 13200 Pulau Pinang, Malaysia
[3]Department of Electrical Engineering, Universiti Malaya, 50603 Kuala Lumpur, Malaysia
[4]Pantai Cancer Institute, Pantai Medical Centre, Bangsar, 50603 Kuala Lumpur, Malaysia
[5]Department of Physics, Universiti Malaya, 50603 Kuala Lumpur, Malaysia
[6]Department of Physics, University of Surrey, Guildford, GU2 7XH Surrey, United Kingdom
[7]Institute of Sciences, Universiti Teknologi MARA, 40450 Shah Alam, Malaysia

E-mail: ahmadtaufek@salam.uitm.edu.my

Abstract: We have investigated the characterisation of new fabricated material Ge doped silica glass thermoluminescence TL dosimeter (Photonic Research Centre, University of Malaya) for medical radiation dosimetry at therapy energy. Previously, the dosimeter has been studied to provide ideal dosimetry system, suitable to ensure an accurate delivery of radiation doses to tumour tissue while minimising the amount of radiation administrated to healthy tissue. Both energies of photon and electron were used in this experiment for a dose range of 1 to 5 Gy. The various sizes of core diameter Ge doped silica glass (120, 241, 362, 483 and 604 μm) were exposed by using linear accelerator at Pantai Medical Centre. For both energies, the optical fibres were found to produce a flat response to a fixed photon and electron doses to within 4% (S.D) of the mean of the TL distribution. In terms of dose response, the fibres provide linear response over the range investigated, from a fraction of 1-5 Gy. The finding shows 120 μm fibres have 1.82 greater dose response than 604 μm fibres irradiated at 6 MV photon with a fixed dose of 3 Gy. While for electron energy 12 MeV, the response shows 120 μm fibres have 1.58 greater dose response compared to 604 μm fibres. The good responses are suitable to make these tailor-made doped silica fibres a promising TL material for use as a dosimetric system in medical radiation therapy.

1. Introduction

Radiation therapy is the use of high energy radiation to shrink tumors and destroy cancer cells. The treatment is delivered by giving high radiation dose to the cancer tissue using a linear accelerator.

However, the dose is also deposited to normal tissue surrounding the cancer and this should be constrained to within acceptable levels.

Desirable characteristics and properties of doped silica glass for use in radiation therapy dosimetry system was studied by [1]. Thermoluminescence properties to be considered in choosing a suitable dosimetric system for radiation therapy dosimetry are linearity between radiation dose and response, sensitive to the signal, TL glow curve, acceptable accuracy and precision, and a good characterized relationship between the dosimeter response in the medium in which dose is to be measured and the calibration radiation field.

Recently, a number of research groups have started to study the potential use of doped silica glass as a radiation TL dosimeter to determine the absorbed dose by patients, in particular overcoming spatial resolution limitations of existing TL dosimetry systems [2]. One of the advantages doped silica glass fibre is its impervious to water and in some instances, like study by [3] about radiation induced attenuation, radioluminescence and optically stimulated luminescence (OSL) [4-6] it has also become possible to locate the fibres within a particular tissue needed [7].

In recent year, the thermoluminescence responses of doped silica glass fibres have been studied for various radiation sources including photon [8-10], electron [9, 11], alpha particles [12], fast neutrons [13] and synchrotron radiation [14]. In all investigation of TL performances of irradiated fibres have shown considerable potential for dosimetric application in radiotherapy, however the study made use of commercial optical fibres available in market which intestinally produce for telecommunication purposes. Therefore, this study will focus on investigation of new tailor-made Ge-doped silica glass fibres produced by Photonic Research Centre, University of Malaya. This study has been made use of 8 mol% of Ge doped silica glass fibre compared to commercial fibres used previously [8-10] which is ~3.1% mol. Literature shows that TL response of doped silica glass fibre relies on dopant concentration at core fibre [8-10].

2. Materials and methods

2.1. Sample preparation
An unclad of 8 % mol Ge doped silica glass with various sizes of core diameter namely 120, 241, 362, 483, 604 μm has been used as the basis of TL dosimetry system. These tailor-made doped silica glass fibres were fabricated at Photonic Research Centre, University of Malaya. The samples were cut with an approximately length of 5.0 ± 0.1 mm by using a cutter. The vacuum tweezer (Dymax 5 Charles Austen Pumps Ltd Surrey, UK) was used to handle the optical fibres. The study involved annealing, sample irradiation, TL measurement and data analysis.

2.2. Annealing process
Annealing process was performed at 400 °C for an hour and cut fibres were placed in alumina ceramic. Annealing process is used to remove any irradiation signal from thermoluminescence material, stabilise the trap structure and reset the dosimeter to initial conditions prior to irradiation. To avoid thermal stress, the fibres were left inside the slowly cooling furnace for a period of 24 hour to finally equilibrate at room temperature. After cooling, the fibre were placed in gelatin capsule, each made to contain 10 pieces of fibre to provide mean value of dose obtain at each irradiation condition. Electronic balance was used to weight each capsule, allowing TL yield to be normalized to unit mass of irradiated fibre (±0.02 mg). The weight of each fibre varied in the range of 0.10-0.35 mg.

2.3. Exposure process
The Ge doped silica glass fibres and TLD-100 were placed at the surface of solid water phantom and were irradiated with 6 MV photon and 12 MeV electron beam at a fixed nominal dose rate of 400 cGy/min using a LINAC (Elekta Synergy ™) at Pantai Hospital Kuala Lumpur, Malaysia. Samples were placed in solid water phantom at D_{max} of 1.5 cm and 2.5 cm depth for both photon energy and

electron energy respectively. The dose delivered by LINAC with field size of 10 cm x 10 cm at SSD of 100 cm.

2.1 Instrumentation

The optical fibre TL yield was readout by a Harshaw 3500 TL reader (Waltham, MA, USA) located at Universiti Teknologi MARA Pulau Pinang, Malaysia. An N_2 atmosphere was used to suppress possible spurious light signals from triboluminescence and also to reduce oxidation of the heating element: preheat temperature is 160 °C for 10s; readout temperature of 300 °C for 13.33s and heating rate cycle of 25 °C s^{-1}. Then, annealing temperature of 300 °C was applied for 10s to erase any thermoluminescence signal. This setting was set to provide an optimal glow curve and free of the effects of superficial traps.

3. Results and discussion

3.1. TL glow curve

Shallow traps, those nearest to the conduction band, are easily emptied at room temperature (RT), leading to measurable fading of the TL signal. The dosimetric traps require somewhat more energy to release trapped electrons, normally forming the peak within which maximum TL yield is obtained and hence are used as the principal peak in dosimetric evaluation. The deep seated traps require appreciable energy in order to be emptied, obtained by high temperature annealing. Figure 1 (a-c) shows the examples of TL glow curve for 120 µm core diameter for increase in absorbed dose from 1 Gy up to 5 Gy.

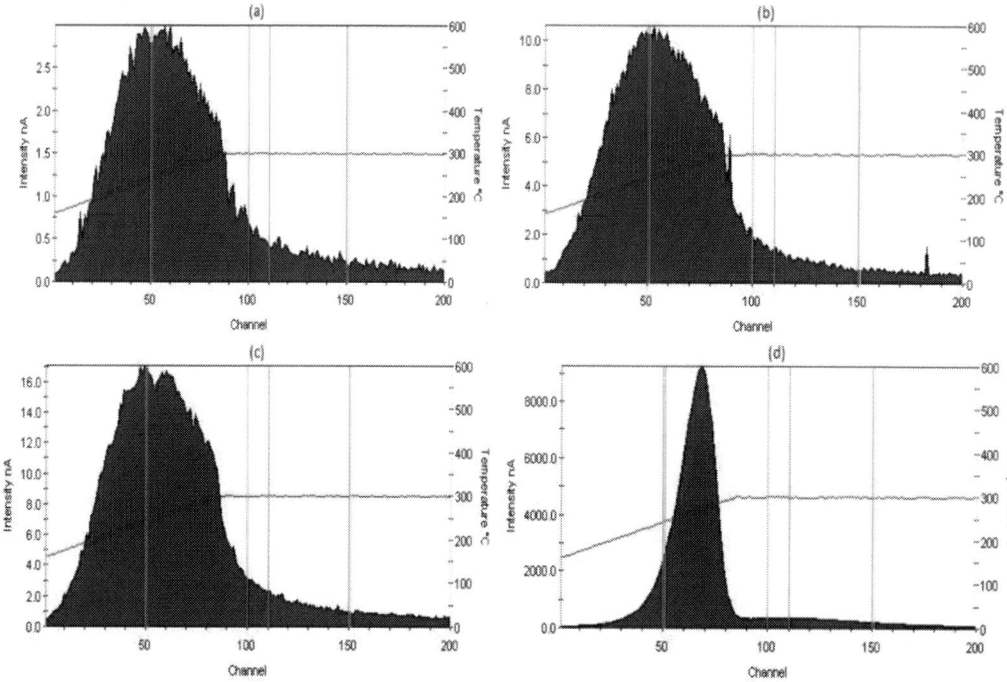

Figure 1. TL Glow Curve of 120 µm for (a) 1 Gy (b) 3 Gy (c) 5 Gy and (d) TL glow curve for TLD-100 irradiated at 3 Gy for 6 MV photon energy.

The maximum peak of TL intensity from 1 Gy to 5 Gy were appeared at approximate channel number 50 and the maximum number of peak TL intensity increase with increasing the irradiation dose. Comparing with the glow curve obtained for TLD-100 at delivered dose of 3 Gy at 6 MV photon irradiation, it shows that the peak number intensity is greater (Figure 1(d)) than fabricated 120 µm core diameter Ge doped silica glass.

3.2. TL response

One of the important characteristic to be a good thermoluminescence dosimeter is linear relationship between TL emission and the absorbed dose. The particular thermoluminescence material gives the great effect to the linearity range. Generally, the response of TL phosphors is linear at low absorbed dose value than becomes supralinear and finally saturates at high values [15]. The fabricated Ge doped silica glass with various sizes of core diameter (120, 241, 362, 483 and 604 µm) were exposed to 6 MV photon energy and 12 MeV electron energy. Figure 2 (a) and Figure 2 (b) displays a linear TL absorbed dose response from 1 Gy up to 5 Gy with respect to sizes of core diameter. The regression (r^2) remains 0.998 or better for both energy irradiations.

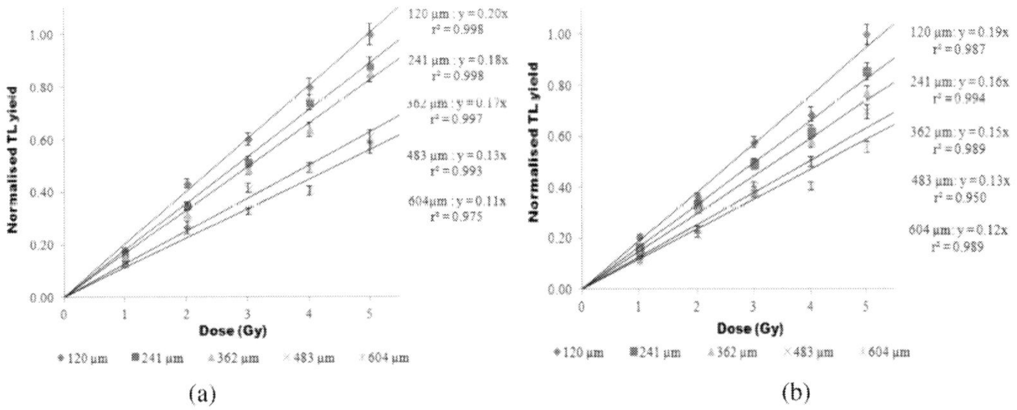

Figure 2. TL response of Ge doped silica glass irradiation of 120, 241, 362, 483, and 604 µm for dose in range 1-5 Gy of (a) 6 MV of photon irradiation energy (b) 12 MeV of electron irradiation energy.

3.3. Energy dependence

To obtain the energy dependence, the sensitivity of Ge doped silica glass was calculated from the dose response relation for different photon and electron energies. The energy response that used in this study is 6 MV photon and 12 MeV electron, the available photon and electron energies for radiotherapy at Pantai Hospital in which the irradiations were carried out. Figure 3 shows that the energy response of 6 MV photon is ~ 5 % greater compared to 12 MeV electron energy. This may be explained by dependence of electron beam scattering and the range of the electron upon density and atomic composition, the effects of heterogeneity upon dose distribution being more pronounced for electron beams than for photons. Electrons being charged particles, energy is lost continuously as they pass through matter and also undergo significant scattering as electrons are light mass particles. The other significant reason is about the effect of air cavities in the loaded gelatin capsule on the dose distribution. It is important to take such factors into account in providing for the precision and accuracy of radiotherapy dosimetry. The need for calibrations carried out under conditions similar to that for field conditions is of great importance [16].

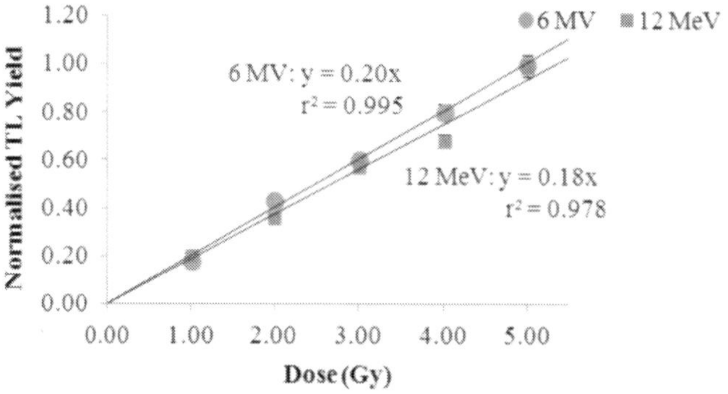

Figure 3. TL response of 120 μm Ge doped silica glass irradiated with 6 MV and 12 MeV versus dose.

3.4. Sensitivity

TL sensitivity can be define [17] as the amount of light released by phosphor per unit of radiation exposure. There are many factors that affect to TL sensitivity dosimeter such as kind and concentration of activators, system of the readout, heating rate, etc. The least square fit of the change in TL response in count per unit mass for 120 μm core fibres irradiated at 6 MV to be 1.82 greater than that obtained for 604 μm core fibres irradiated at same photon energy (Figure 2 (a)). While for electron energy, the least square fit of the change in TL response in count per unit mass for 120 μm core fibres irradiated at 12 MeV to be 1.58 greater than that obtained for 604 μm core fibres irradiated at same energy as shown in Figure 2 (b). Therefore, the average ratios finds that 120 μm core diameter Ge doped silica glass has better response and is more sensitive compared to 604 μm core diameter Ge doped silica glass for both energy irradiations.

4. Conclusion

At 6 MV and 12 MeV, the tailor-made Ge doped silica glass fibres fabricated by Photonic Research Centre, University of Malaya, provide good characteristic dosimeter that used in medical treatment in term of linearity, dose response, energy dependence and sensitivity point to favorable applications within the dose ranges studied. The Ge doped silica glass fibres is an attractive viable dosimeter in radiotherapy due to its small physical size, flexibility and low cost. Further study on its application is significant to establish this tailor-made Ge doped silica glass fibres for potential use in medical treatment applications to be alternative TL dosimeter available.

Acknowledgements

The authors would like to thank the Universiti Teknologi MARA (UiTM) for provision of a grant under the Excellent Research Fund Scheme, Grant 600-RMU/ST/DANA 5/3/DST (11/2012). We also acknowledge our gratitude for receipt of a University of Malaya-Ministry of Education UM-MOE High Impact Research Grant UM.C/625/1/HIR/33 to support part of this project. We also would like to thank the Ministry of Education Malaysia (MOE) for supporting a M.Sc. research studentship for Nurul Arina Zahaimi.

References

[1] Zahaimi, N.A., et al., Dopant concentration and thermoluminescence (TL) properties of tailor-made Ge-doped SiO2 fibres. *Radiation Physics and Chemistry*, 2014. **104**: p.297–301.

[2] Hashim, S., et al., The thermoluminescence response of doped SiO2 optical fibres subjected to photon and electron irradiations. *Appl Radiat Isot*, 2009. **67**(3): p. 423-7.

[3] Huston, A.L., et al., Remote optical fiber dosimetry. *Nuclear Instruments and Methods in Physics Research Section B: Beam Interactions with Materials and Atoms*, 2001. **184**(1–2): p. 55-67.

[4] Aznar, M.C., et al., Real-time optical-fibre luminescence dosimetry for radiotherapy: physical characteristics and applications in photon beams. *Phys Med Biol*, 2004. **49**(9): p. 1655-69.

[5] Justus, B., et al., Optically stimulated luminescence radiation dosimetry using doped silica glass. *Radiation Protection Dosimetry*, 1997. **74**(3): p. 151-154.

[6] Benevides, L.A., et al., Characterization of a fiber-optic-coupled radioluminescent detector for application in the mammography energy range. *Med Phys*, 2007. **34**(6): p. 2220-7.

[7] Yusoff, A.L., R.P. Hugtenburg, and D.A. Bradley, Review of development of a silica-based thermoluminescence dosimeter. *Radiation Physics and Chemistry*, 2005. **74**(6): p. 459-481.

[8] Abdulla, Y.A., Y.M. Amin, and D.A. Bradley, The thermoluminescence response of Ge-doped optical fibre subjected to photon irradiation. *Radiation Physics and Chemistry*, 2001. **61**(3–6): p. 409-410.

[9] Abdul Rahman, A.T., A. Nisbet, and D.A. Bradley, Dose-rate and the reciprocity law: TL response of Ge-doped SiO2 optical fibers at therapeutic radiation doses. *Nuclear Instruments and Methods in Physics Research Section A: Accelerators, Spectrometers, Detectors and Associated Equipment*, 2011. **652**(1): p. 891-895.

[10] Issa, F., et al., Ge-doped optical fibres as thermoluminescence dosimeters for kilovoltage X-ray therapy irradiations. *Nuclear Instruments and Methods in Physics Research Section A: Accelerators, Spectrometers, Detectors and Associated Equipment*, 2011. **652**(1): p. 834-837.

[11] Hashim, S., et al., The thermoluminescence response of doped SiO2 optical fibres subjected to photon and electron irradiations. *Applied Radiation and Isotopes*, 2009. **67**(3): p. 423-427.

[12] Ramli, A.T., et al., The thermoluminescence response of doped SiO2 optical fibres subjected to alpha-particle irradiation. *Applied Radiation and Isotopes*, 2009. **67**(3): p. 428-432.

[13] Hashim, S., et al., The thermoluminescence response of doped SiO optical fibres subjected to fast neutrons. *Applied Radiation and Isotopes*, 2010. **68**(4): p. 700-703.

[14] Abdul Rahman, A.T., et al., The thermoluminescence response of Ge-doped silica fibres for synchrotron microbeam radiation therapy dosimetry. *Nuclear Instruments and Methods in Physics Research Section A: Accelerators, Spectrometers, Detectors and Associated Equipment*. **619**(1–3): p. 167-170.

[15] Yaakob, N.H., et al., Electron irradiation response on Ge and Al-doped SiO2 optical fibres. *Nuclear Instruments and Methods in Physics Research Section A: Accelerators, Spectrometers, Detectors and Associated Equipment*, 2011. **637**(1): p. 185-189.

[16] Bradley, D.A., et al., Review of doped silica glass optical fibre: Their TL properties and potential applications in radiation therapy dosimetry. *Applied Radiation and Isotopes*, 2012. **71, Supplement**(0): p. 2-11.

[17] McKeever, S.W.S. and P.D. Townsend, Thermoluminescence dosimetry materials: properties and uses. 1995: *Nuclear Technology Pub.*

Contrast enhancement of magnetic resonance imaging (MRI) of polymer gel dosimeter by adding Platinum nano- particles

N Deyhimihaghighi[1], N Mohd Noor[2], N Soltani[1], R Jorfi[1] M Erfani Haghir[1], M Z Adenan[1], E Saion[1] and M U Khandaker[3]

[1] Department of Physics, Faculty of Science, Universiti Putra Malaysia, Serdang, Selangor, 43400, Malaysia
[2] Department of Imaging, Faculty of Medicine and Health Sciences, Universiti Putra Malaysia, 43400 Serdang, Selangor, Malaysia.
[3] Department of Physics, University of Malaya, 50603 Kuala Lumpur, Malaysia

E-mail: parissaday@gmail.com

Abstract. In this research different concentrations of Platinum nanoparticles (Pt-NPs) embedded in PAGAT polymer gel and irradiated with Cobalt -60 (^{60}Co). The OD of samples indicate significant dose enhancement caused by Pt-NPs. Different concentrations of Pt-NPs were investigated: 0.5×10^{-2},1×10^{-2}, 2×10^{-2} and 3×10^{-2} mg/l and these were presented the enhancement in OD by the factor of 11.56, 27.10, 15.84 and 10.07 respectively. As the mean energy of ^{60}Co is 1.25 MeV and atomic number (Z) of Pt is 78, the predominant effect will be Compton effect and Photoelectric cannot occurred. This work studied the feasibility of using PAGAT polymer gel combined with Pt-NPs as a suitable tool to perform dosimetric investigation of nanoparticles application in radiotherapy as dosimeter and dose enhancer.

1. Introduction

One of the primary aims of radiation therapy is to reduce the dose to healthy tissue and increase the dose to tumor cells. Developing technology help us to reach this goal by using metal nanoparticles application in field of radiology and radiation therapy. Experiments in living cells and *in vivo* indicate the effectiveness of the metal nanoparticles when apply low energy x-ray radiations (below conventional 1 MeV linear accelerator (linac) radiation. More studies on DNA have been carried out in order to better understand the fundamental processes of sensitization and to further improve of current method. It is proposed a new method based on the combination of Platinum nanoparticles (Pt-NPs) with irradiation by fast ions (hadron therapy)[1]. It was observed that nanoparticles could enhance strongly lethal damage in DNA. It was shown that the sensitization in presence of nanoparticles is increased because of auto-amplified electronic cascades inside the nanoparticles, which reinforces the energy deposition in the vicinity of the metal. Therefore, the combination of fast ion radiation (hadron therapy) with platinum nanoparticles could improve cancer therapy protocols effectively.

Subsequently, Porcel [2] tested different rang of x-ray energy combine with Pt-NPs to investigate the effect on the DNA. Obviously photoelectric was the predominant effect in this study and produced photoelectrons ionized the Platinum atoms. These findings showed that any high Z nanoparticles might behave like radiation enhancer, no matter ionizing radiation is electromagnetic or charged particles. Enhancement the dose by high Z materials depend on source of irradiation energy[3]. In dosimetric point of view, in kilovoltage range, photoelectric will be the predominant effect, as photoelectric depend upon Z^3/E^3 (Z=atomic number and E= energy) [4]. But it is need to investigate the Megavoltage ranges of energy like gamma ray (γ-ray) by ^{60}Co when the Compton effect is predominant. In this work different concentration of Pt-NPs were embedded in PAGAT polymer gel and irradiated by γ-ray to investigate the dose enhancement in MeV range of energy, which the predominant effect is Compton effect.

2. Materials and methods

2.1. Platinum nanoparticles synthesize

The beam from Nd:YAG laser was focused by a lens (F = 25 mm) on the surface of a platinum plate (99.99%, thickness 1.0 mm) fixed inside a small container with 6 ml of distilled water. The laser pulse duration was about 10 ns, the pulse-repetition rate 10 Hz, and the maximum energy of the pulse is 0.34 J/pulse. The duration of the ablation experiment was 1 hour. It has been demonstrated that this physical method is free of any reducing agents and is applicable to the preparation of nanoparticles of any metal elements through ready formation of the metal atoms by laser ablation [5].

Immediately after the ablation experiment, the UV-Vis (Shimadzu UV-1650PC) absorption spectra of Platinum colloids were measured. The peak in 216 nm showed that the Pt-NPs was formed in water. A transmission electron microscope (TEM) (Hitachi H-800) was employed to take electron microphotographs of the resultant nanoparticles[6]. The average size for formed Pt-NPs was 10 nm. Atomic absorption (AA) was used for detecting the concentration of metals in solution, which was 1 mg/l.

2.2. Nanoparticle incorporation in gel

PAGAT (Acrylamide, gelatin, Hydroquinone, THPC)[7] was prepared at 38 °C. Pt-NPs solution was added in the last step of preparation to the PAGAT. Two groups of gels were prepared, first without nanoparticles and second with nanoparticles. The sample not containing Pt-NPs served as control. Gel homogenization was achieved by routine mechanical mixed. Four different Pt-NPs concentrations were considered: 0.5×10^{-2}, 1×10^{-2}, 2×10^{-2} and 3×10^{-2} mg/l.

2.3. Irradiation procedure and MRI reading

Gel samples were irradiated by ^{60}CO at University Kebangsaan Malaysia (UKM) using gamma cell (220 Excel model 220 E S/N 117R) at the dose rate of 2.2550 kGy/h. The samples were placed inside the water to guarantee backscattering conditions and created the tissue equivalent situation during irradiation. Control gel samples and samples containing Pt-NP were irradiated with a dose 6 to 25Gy with interval 2 Gy.

Optical density measurement was carried out by densitometer (Vectoreen, model 07-440, USA), which correlates with received dose and contrast in printed MRI images. Samples were scanned by Magnetic Resonance Imaging (MRI) (Magneton SP Siemens, Germany) 12 hours after irradiation. A multiple spin-echo sequence was used with the following parameters: *T2* weighted imaging was carried out using 64 spin-echoes with a TE of 74 ms, TR of 5000 ms, slice thickness of 4 mm, matrix size of 256 × 256, FOV 250 mm [7, 8].

3. Results and discussion

PAGAT polymer gel response to γ-ray was characterized by comparing the OD in samples without nanoparticles and with Pt-NPs. The PAGAT response was found to be linear over the entire dose range explored from 6 Gy up to 25 Gy. Here also we can conclude that maximum enhancement was seen in PAGAT doped with 1×10^{-2} mg/l Pt-NPs which is seen in Figure 1, (b).

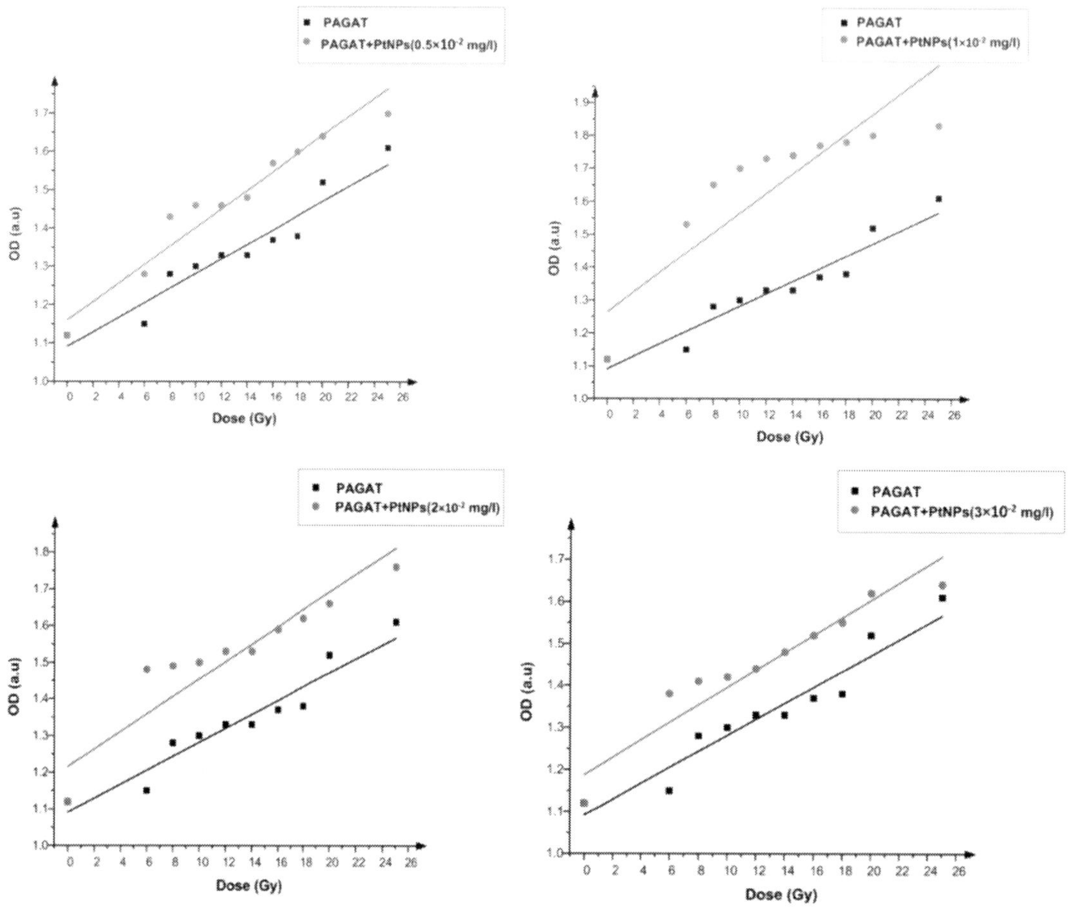

Figure 1. Optical density versus dose without nanoparticles and with different concentration of Pt-NPs (a) 0.5×10^{-2} mg/l, (b) 1×10^{-2} mg/l, (c) 2×10^{-2} mg/l, (d) 3×10^{-2} mg/l.

The OD of non-irradiated samples (with and without Pt-NPs) were evaluated. Chemical interactions between Pt-NPs and gel were discarded considering the OD variation found was 0. Table 1 presents the respective percentage of enhancement from samples non-irradiated (0 Gy) to show no chemical interaction.

The dose enhancement due to Pt-NPs presence in the tissue-equivalent gel was quantified comparing the OD of all different concentrations of Pt-NPs irradiated with the different dose (6 to 25 Gy). According to Figure 1, Concentration (a) presents an increase of 11.16% in OD comparing to gel without Pt-NP; concentrations (b), (c) and (d) present increases of 27.10%, 18.54% and 10.07% respectively (see Table 1).

According to graph which Evans [9] sketch, in the case of γ-ray as the K-edge of Pt is 78.39 KeV and the mean energy of ^{60}Co is 1.25 MeV, then the photoelectric is not occurred, but in presence of Pt-NPs inside, the predominant effect is Compton. As we know the Compton effect occur with free electrons, then the possibility of Compton effect is not related to atomic number of material but to the number of electron per gram. The number of electron per gram is the same for all material except Hydrogen, then what makes a difference between various materials is the number of electron per cubic centimeters (e/cm^3)[4]. The number of e/cm^3 for Pt was calculated (5.17×10^{24}), which was more than other material like gold (4.66×10^{24}) and silver (2.75×10^{24}) and also muscle (3.36×10^{23}).

Table 1. Percentage of enhancements with different concentration of Pt-NPs.

Dose (Gy)	Concentration of Pt-NPs (mg/l)			
	0.5×10^{-2}	1×10^{-2}	2×10^{-2}	3×10^{-2}
0	0	0	0	0
6	11.30	33.04	28.70	20
8	11.72	28.91	16.41	10.16
10	12.31	30.77	15.38	9.23
12	9.77	30.08	15.04	8.27
14	11.28	30.83	15.04	11.28
16	14.60	29.20	16.06	10.95
18	15.94	28.99	17.39	12.32
20	7.89	18.42	9.21	6.58
25	5.59	13.66	9.32	1.86
Average (%)	11.16	27.10	15.84	10.07

It indicated the fact that with fewer amount of Pt nanoparticles (0.5×10^{-2} mg/l) the OD was increasing compare to PAGAT without nanoparticles and this trend continue until 1×10^{-2} mg/l Pt-NPs. This was due to Compton electrons produced, which can increase the probable of radiolysis of water and arise the amount of free radicals, produced. These free radicals can polymerize more monomers and make the gel more opaque.

But the question was that why 1×10^{-2} mg/l Pt-NPs showed better enhancement than 2×10^{-2} and 3×10^{-2} mg/l. Ertl *et al.* [10] explained that iodine as one of dose enhancer was scavenger by radiologically forming $OH^{\bullet}$ radicals and as the reaction rate of I^{-} with $OH^{\bullet}$ and monomer with $OH^{\bullet}$ were approximately the same, then there was a competition between the monomers and the halide ions to react with $OH^{\bullet}$ resulting in less initiation of polymerization. And the result was to decrease the dose response and the decrease was more pronounced in higher concentration of dose enhancer [11]. In current research, it can also be explained in the same way. If the Pt-NPs, which have positively charge surface, like other nanoparticles, interact with $H^{\bullet}$ and $OH^{\bullet}$, it will reduce the initiation of polymerization, as free radicals are necessary to initiate the polymerization process. Then the optimum concentration of Pt-NPs in the volume of PAGAT that I made was 1×10^{-2} mg/l. Therefore, the application of Pt-NPs as a dose enhancer or dosimeter was proved.

Dose enhancements verified by gel dosimetry presented in this work instigate future investigation. Different size of nanoparticles can be investigated using polymer gel dosimetry aiming to optimize the application of nanoparticles in radiation therapy.

4. Conclusion

The feasibility of using PAGAT polymer gel combine with Pt-NPs as dose enhancer or dosimeter to increase the OD in MRI was investigated in this work. The result showed that in presence of 1×10^{-2} mg/l Pt-NPs, the OD increased by factor of 27.10%.

References

[1] Porcel, E., et al. Photosensitization of plasmid-DNA loaded with platinum nano-particles and irradiated by low energy X-rays. in Journal of Physics: Conference Series. 2011. IOP Publishing.

[2] Porcel, E., et al., Platinum nanoparticles: a promising material for future cancer therapy? Nanotechnology, 2010. 21(8): p. 085103.

[3] McMahon, S.J., et al., Radiotherapy in the presence of contrast agents: a general figure of merit and its application to gold nanoparticles. Physics in medicine and biology, 2008. 53(20): p. 5635.

[4] Khan, F.M., The physics of radiation therapy. Vol. 3. 2003: Lippincott Williams & Wilkins Philadelphia.
[5] Mafune, F., et al., Formation of stable platinum nanoparticles by laser ablation in water. The Journal of Physical Chemistry B, 2003. 107(18): p. 4218-4223.
[6] Pyatenko, A., et al., Synthesis of silver nanoparticles by laser ablation in pure water. Applied Physics A, 2004. 79(4-6): p. 803-806.
[7] Venning, A., et al., Investigation of the PAGAT polymer gel dosimeter using magnetic resonance imaging. Physics in medicine and biology, 2005. 50(16): p. 3875.
[8] Rahman, W.N., et al., Dosimetry And Its Enhancement Using Gold Nanoparticles In Synchrotron Based Microbeam And Stereotactic Radiosurgery, in 6th International Conference on Medical Applications of Synchrotron Radiation, K.K.W. Siu, Editor. 2010. p. 107-110.
[9] Evans, R.D., Gamma rays. Chap. 8e in "American Institute of Physics Handbook", 3rd Ed., McGraw–Hill, 1972.
[10] Ertl, H.H., L.E. Feinendegen, and H. Heiniger, Iodine-125, a tracer in cell biology: physical properties and biological aspects. Physics in medicine and biology, 1970. 15(3): p. 447.
[11] Meesat, R., et al., Evaluation of the dose enhancement of iodinated compounds by polyacrylamide gel dosimetry. Physics in medicine and biology, 2009. 54(19): p. 5909.

Journal of Physics: Conference Series **546** (2014) 012014 doi:10.1088/1742-6596/546/1/012014

Multi-directional radiation detector using photographic film

L K Junet, Z A Abdul Majid, A H Sapuan, I S Sayed and N F Pauzi

Kulliyyah of Allied Health Sciences, International Islamic University Malaysia, Kuantan, Pahang, 25200, Malaysia

E-mail: zafriazran@yahoo.com

Abstract: Ionising radiation has always been part of our surrounding and people are continuously exposed to it. Ionising radiation is harmful to human health, thus it is vital to monitor the radiation. To monitor radiation, there are three main points that should be observed cautiously, which are energy, quantity, and direction of the radiation sources. A three dimensional (3D) dosimeter is an example of a radiation detector that provide these three main points. This dosimeter is able to record the radiation dose distribution in 3D. Applying the concept of dose detection distribution, study has been done to design a multi-directional radiation detector of different filter thicknesses. This is obtained by designing a cylinder shaped aluminum filter with several layers of different thickness. Black and white photographic material is used as a radiation-sensitive material and a PVC material has been used as the enclosure. The device is then exposed to a radiation source with different exposure factors. For exposure factor 70 kVp, 16 mAs; the results have shown that optical density (OD) value at 135^0 is 1.86 higher compared with an OD value at 315^0 which is 0.71 as the 135^0 area received more radiation compare to 315^0 region. Furthermore, with an evidence of different angle of film give different value of OD shows that this device has a multidirectional ability. Materials used to develop this device are widely available in the market, thus reducing the cost of development and making it suitable for commercialisation.

1. Introduction

Ionising radiation has always been part of our surrounding and people are continuously exposed to it. This radiation is emitted by natural and/or man-made sources [1]. In the medical community, ionising radiation is used for two purposes; diagnosis and therapy [2]. As radiologic diagnosis provides greater accuracy for disease interpretation, the use of ionising radiation in medical application has rapidly increased [3].

Ionising radiation is an invisible form of energy that can cause harm to human health with cancer being the most popular effect [3,4]. In 1986, the reactor explosion in Chernobyl had caused a commotion regarding the effect of radiation to the human health. In the disaster, the explosion of radioactive substances and fire lasted for ten days and caused damage to the active zone of the reactor. This led to large quantities of radioactive particles being spread into the atmosphere, and the area was contaminated [5]. It was reported [6] that thyroid cancer cases had increased among the Chernobyl's victims fourteen years after the accident.

To evaluate the radiological relevant doses received by the community and the potential impact of ionising radiation on human health, it is necessary to observe the radiation doses received [1]. For this reason, radiation monitoring devices or dosimeters is required. In order to monitor the radiation, there are three main points that should be observed cautiously which are energy, quantity and direction of the radiation sources. A three dimensional (3D) dosimeter is an example of radiation detector that covers these three important aspects. This dosimeter is made of radiation sensitive materials (or chemicals) and is able to record the radiation dose distribution in 3D [7]. Applying the concept of dose detection distribution, study has been done to design a multi-directional radiation detector consisting of different filter thicknesses. Black and white photographic film is chosen to be used as a medium of radiation detector in this study.

Generally, photographic film has been used in photographic dosimeters such as film badges. This type of dosimeter is used for monitoring cumulative exposure of radiations such as X-rays and gamma rays [8] without showing the direction of the radiation sources. According to Dowsett, et al. [9] the major advantages of this dosimeter are that it provides a permanent record, able to distinguish different energies of photons and it can measure different types of ionising radiation. Photographic dosimeter makes use of film as a radiation-sensitive material. This film contains a special emulsion layer that can detect radiation and is later analysed to measure the doses [10]. Lalos [11] stated that for photographic dosimeter, the type of film used to detect radiation was basically similar to ordinary black and white photographic film. Furthermore, photographic films also contain a special emulsion layer and because of that, it can be used to detect ionising radiation [12]. Once the film is developed, exposed area forms an optical density (OD) in response to incident radiation. This OD represents the quantity of ionising radiation exposed to the film and it can be read using a densitometer. This type of film is most sensitive to ionising radiations that have energy of 50 keV and above [3].

To identify ionising radiations' penetrating power, filters are used. With filtration it is possible to determine the different radiation energy. Filtration is the process of shaping the x-ray beam or eliminating the low energy photons by placing an absorbing material along the primary beam. This absorbing material also known as filter is a sheet of metal that is designed to selectively absorb photons from the ionising radiation [10]. Selecting proper filter material and thickness is important because different types of filter material and thickness may results in different film exposure [12] since each material has different atomic number. The higher the atomic number, the lesser the amount of radiation that can penetrate the material to be detected by the film. Aluminium which has atomic number of 13, was selected as the best material to be used as filter for this research.

2. Materials and methods

2.1. Materials

2.1.1 Construction of Multi-directional Radiation Detector. The multi-directional radiation detector consists of two main parts which are film casing and radiation sensitive material. The film casing is composed of polyvinyl chloride (PVC), aluminum middle rod and aluminum filters with several layers. A schematic diagram of the multi-directional radiation detector is shown in Figure 1. For radiation sensitive material, ILFORD DELTA 3200 Professional (HARMAN technology ltd, Cheshire, England) black and white photographic film was used. Aluminum middle rod is the innermost component, followed by photographic film with an effective area (area that receives ionising radiation) of 25mm, PVC enclosure and aluminum filter with several layers as the outermost component.

The detector is designed as a portable device with cylindrical shape particularly because this shape embraced similar surface on every angle. Thus, enlarging the surface for ionising radiation reaching the film. PVC was picked for the enclosure as it is a material opaque to ionising radiation; hence, there is no possibility for the film to fog up due to light.

The aluminum filter is comprised of four layers of different thicknesses; 0.75mm, 1.50mm, 2.25mm and 3.00mm. A schematic diagram of the aluminum filter is shown in Figure 2. The purpose of these layers of filter is to determine the radiation penetration energy and the quantity of radiation reaching the film. The purpose of aluminum middle rod is to prevent the ionising radiation from

Journal of Physics: Conference Series **546** (2014) 012014 doi:10.1088/1742-6596/546/1/012014

reaching the film on more than one side which might lead to inaccurate result. Furthermore, the use of black and white photographic film as radiation sensitive material was due to its special emulsion layer

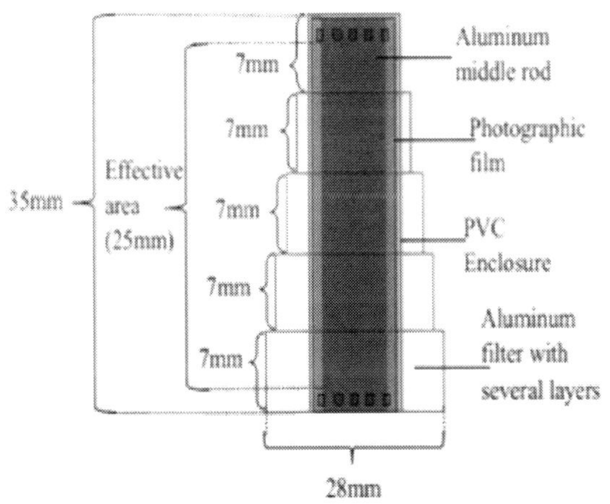

Figure 1. Schematic diagram of the radiation detector.

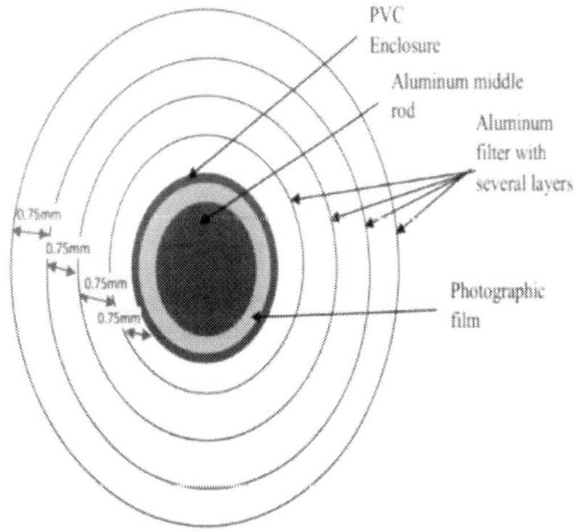

Figure 2. Schematic diagram of the thickness layers for aluminum filter.

that can be used to detect the ionising radiation [12].

2.2 Methods

This research was done in Diagnostic Imaging and Radiotherapy Department at International Islamic University Malaysia (IIUM), Kuantan, Pahang, Malaysia. Siemens POLYDOROS IT 80 x-ray machine was used as ionising radiation source. Before exposure, photographic film of 60mm in length was inserted in the space between PVC enclosure and aluminum middle rod. This step was done in a darkroom in order to avoid film's fog that may cause imprecise result.

The detector was exposed to ionising radiation with fixed kilovoltage peak (kVp); 70 kVp and different milliampere-seconds (mAs). 70 kVp was chosen as the optimum parameter to detect the effect of different aluminium filter thicknesses in relation to ionising radiation exposure. During this procedure, only one area was exposed directly to ionising radiation which is approximately 45^0 to 225^0 from reference angle (0^0) of the detector. The field size used was 8mm x 15mm. A schematic diagram of this procedure is shown in Figure 3. The purpose of this technique is to identify the differences in OD value between direct and indirect exposed region.

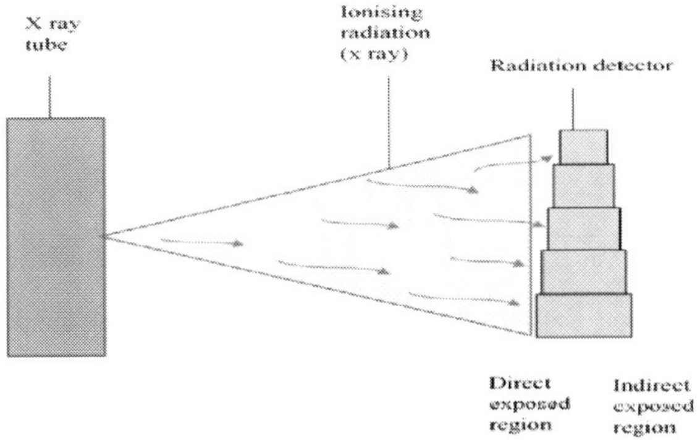

Figure 3. Schematic diagram of multi-directional radiation detector exposed to ionising radiation.

Then, the film was processed using a procedure recommended by ILFORD [13,14]. The processed photographic film was divided into 17 horizontal segments of 3mm each and 5 vertical segments of 7mm each. The OD values were then taken at the center of each point. A densitometer (Hand Held Deluxe Digital Clamshell Densitometer ST07443) was used to measure the OD. The control OD value was measured on film that was totally unexposed to ionising radiation.

3. Results and discussions

Experimental results of this research are shown in Figure 4 and 5. The control OD value for this photographic film is 0.46.

Compared the control OD value and the results gathered for each exposure, there is significant increase in OD when photographic film is exposed to ionising radiation. For example, when a photographic film was exposed to 16 mAs (Figure 4) radiation with 3.00 mm aluminium filter thickness, the OD value for 0^0 (indirect exposed region)and 135^0 (direct exposed region) increased to 0.71 and 1.44 respectively from the control value of 0.46.

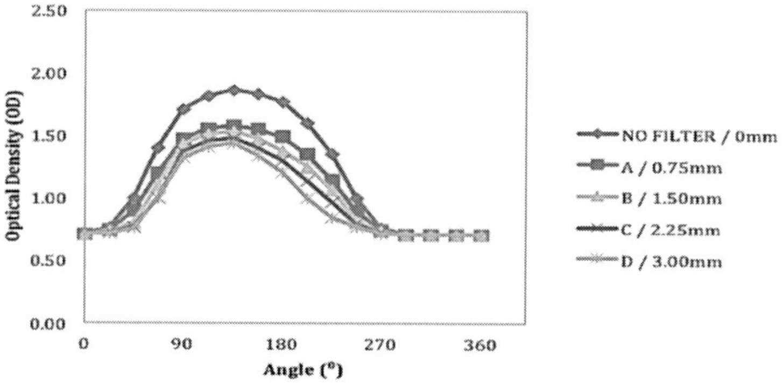

Figure 4. OD values for different angle of radiation detector vs filter thickness for 16 mAs.

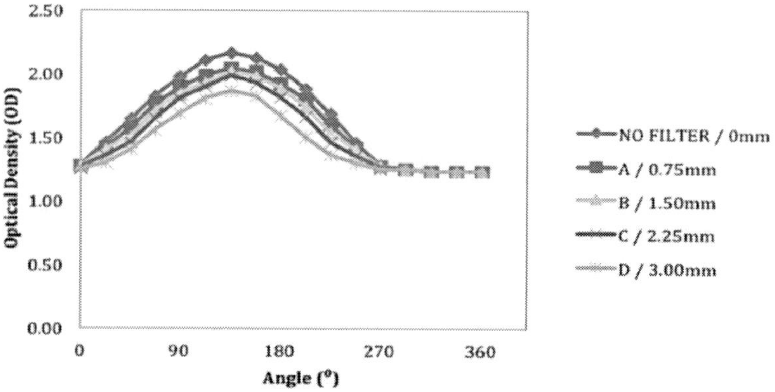

Figure 5. OD values for different angle of radiation detector vs filter thickness for 36 mAs.

The black and white photographic film was chosen as radiation sensitive material, because as mentioned by Lalos [11], the film used in photographic dosimeters is essentially similar to ordinary black and white photographic film. Furthermore, according to Majid et al. [12] black and white photographic film also contains special emulsion layer and can be used to detect ionising radiation.

This special emulsion layer is composed of silver halide crystals that produce OD when exposed to ionising radiation [8]. The figures shows the relationship between OD and filter thickness, proved that black and white photographic film can be used as a radiation-sensitive material. Additional advantages for using films as radiation-sensitive material are its ability to distinguish different photons' energy [9], provide permanent record [11] and cheap as it is a low cost material that is easily available [12].

Both figures show the relationship between OD values for angles of the multi-directional radiation detector with a given mAs. In this research the kVp value was kept constant at 70 kVp for each exposure but different values of mAs were used. This is because, as stated by Fauber [15], mAs controls the quantity of ionising radiation reaching the film.

For 0 mm filter, Figure 4 shows that OD value at 135^0 is 1.86 and Figure 5 shows OD value for 36 mAs at 135^0 is 2.16. These results show that OD value for 36 mAs is higher than 16 mAs. The higher the mAs value, the more ionising radiation reached the film, hence increased the OD value [9].

In this research, ionising radiation are detected based on filter thickness and angle of the multi-directional radiation detector. As referred in Figure 4, the same quantity of ionising radiation applied to the 90^0 region of the multi-directional radiation detector produces different OD values. This is because the radiation penetrates various filter thickness before reaching the film. For this region (90^0) in Figure 4, OD values recorded are 1.71, 1.47, 1.43, 1.37 and 1.32 at 0mm, 0.75mm, 1.5mm, 2.25mm and 3.0mm filter thickness respectively. This shows that increase in filter thickness will decrease amount of ionising radiation reaching the film, hence decreasing the OD value. As mentioned by Majid et al. [12] different material thickness may produce different amount of radiation reaching the film. Thus, this can determine the different radiation energy.

The value of OD is not similar at every angle of multi-directional radiation detector. For example in Figure 4, for a 3.00 mm filter thickness the results shows OD value at 135^0 is 1.44 which is higher compared to OD value of 0.72 at 270^0. This is because the area at 135^0 received more ionising radiation relative to 270^0 region; only one area was exposed directly to ionising radiation (direct exposed region). This technique was applied to identify the difference of OD value between the directly exposed region and indirectly exposed region. Results show difference in the OD values, hence this technique is practiced.

There is some limitations when using photographic film which are it cannot provide a real time dose information and very sensitive to high temperature [8]. High temperatures or excess humidity to photographic film will cause fog resulting in inaccurate exposure reading [3].

4. Conclusion

This study was found that this device has a multi-directional ability of detection of radiation from 0^0 to 360^0 ,supported by evidence of different angles of film provide different values of OD. The results of penetration depend on the thickness of the aluminum filter as different thicknesses will affect ionising radiation penetration capability. Materials used to develop this device are widely available in the market, thus reducing the cost of development and making it suitable for commercialisation.

Acknowledgement

The author would like to thank the Kulliyyah of Allied Health Sciences and International Islamic University Malaysia (IIUM) Endowment Fund for financial assistance for this research project.

References

[1] Pröhl G 2003 Radioactivity in the terrestrial environment *Radioactivity in the Environment* **4** ed Scott E M 87-90

[2] Graham D T, Cloke P & Vosper M 2007 *Principles of Radiological Physics* 5th edition (Philadelphia: Elsevier) p 9

[3] Sherer M A S, Visconti P J, & Ritenour E R 2011 *Radiation Protection in Medical Radiography* 6th edition (Missouri: Mosby Elsevier) pp 11 & 78.

[4] Davis S 2012 Health risks associated with environmental radiation exposures *Journal of Radiological Protection*. **32** N21-N25

[5] Kortov V & Ustyantsev Y 2013 Chernobyl accident: Causes, consequences and problems of radiation measurements *Radiation Measurements* **55** 12–16

[6] United Nation Scientific Committee on the Effects of Atomic Radiation (UNSCEAR) 2000 *Exposures and effects of the Chernobyl accident* Annex J.

[7] Baldock C, Deene Y D, Doran S, Ibbott G, Jirasek A, Lepage M, McAuley K B, Oldham M & Schreiner L J 2010 Polymer Gel Dosimetry *Phys. Med. Biol.* **55** (5) R1-R63

[8] Attix F H 2004 *Introduction to Radiological Physics and Radiation Dosimetry* (Weinheim: Wiley VCH) pp 411-417

[9] Dowsett D J, Kenny P A & Johnston R E 2006 *The Physics of Diagnostic Imaging* 2nd edition (New York: Hodder Arnold) p 65

[10] Carlton R R. & Adler A M 2006 *Principles of Radiographic Imaging* 4th edition (New York: Thomson Delmar Learning) pp 141, 164-171

[11] Lalos G 1989 Film Badge Dosimetry in Atmospheric Nuclear Test *National Academy Press* **4** 35

[12] Majid Z A A, Junet L K, Hazali N, Abdullah A A & Hanafiah M A K M 2013 Variable filtered photographic film as a radiation detector for environmental radiation monitoring *AIP Conf. Proc.* **1528** 123-128

[13] ILFORD Information Sheet 2003 *Processing Your First Black and White Film* HARMAN technology limited

[14] ILFORD Film Processing Chart 2010 HARMAN technology limited

[15] Fauber T L 2009 *Radiographic Imaging and Exposure* 3rd edition (Missouri: Mosby Elsevier) pp 32-33

Assessment of radiation safety awareness among nuclear medicine nurses: a pilot study

N A Yunus[1], M H R O Abdullah[2], M A Said[3] and P E Ch'ng[4]

[1] Faculty of Applied Sciences, Universiti Teknologi MARA, Shah Alam, Selangor, 40450, Malaysia
[2] Department of Applied Sciences, Universiti Teknologi MARA, Permatang Pauh, Pulau Pinang, 13500, Malaysia
[3] Department of Nuclear Medicine, National Cancer Institute, Putrajaya, 62250, Malaysia
[4] Department of Computer and Mathematical Sciences, Universiti Teknologi MARA, Permatang Pauh, Pulau Pinang, 13500, Malaysia

E-mail: norazwa0608@yahoo.com

Abstract. All nuclear medicine nurses need to have some knowledge and awareness on radiation safety. At present, there is no study to address this issue in Malaysia. The aims of this study were (1) to determine the level of knowledge and awareness on radiation safety among nuclear medicine nurses at Putrajaya Hospital in Malaysia and (2) to assess the effectiveness of a training program provided by the hospital to increase the knowledge and awareness of the nuclear medicine nurses. A total of 27 respondents attending a training program on radiation safety were asked to complete a questionnaire. The questionnaire consists 16 items and were categorized into two main areas, namely general radiation knowledge and radiation safety. Survey data were collected before and after the training and were analyzed using descriptive statistics and paired sample t-test. Respondents were scored out of a total of 16 marks with 8 marks for each area. The findings showed that the range of total scores obtained by the nuclear medicine nurses before and after the training were 6-14 (with a mean score of 11.19) and 13-16 marks (with a mean score of 14.85), respectively. Findings also revealed that the mean score for the area of general radiation knowledge (7.59) was higher than that of the radiation safety (7.26). Currently, the knowledge and awareness on radiation safety among the nuclear medicine nurses are at the moderate level. It is recommended that a national study be conducted to assess and increase the level of knowledge and awareness among all nuclear medicine nurses in Malaysia.

1. Introduction

Radiological imaging is an essential tool to medical staffs in diagnosing diseases. Among the common imaging modalities used in diagnostic are radiography (x-rays), fluoroscopy, mammography, computed tomography (CT) and positron emission tomography (PET). The utilization of ionizing radiation in these tools have caused potential risk to patients [1].

Ionizing radiation has hazardous effects on biological systems [2, 3]. There are two types of radiation effect on human health namely the stochastic effect and the deterministic effects. Stochastic effect is an independent-dose effect and can cause illness such as cancer [4, 5]. Dependent-dose effect are called deterministic effects and the effects are immediate such as radiation burn and acute radiation syndrome [5,6]. Despite the fact that ionizing radiation causes hazardous effects on human health, high-energy radiation such as x-ray and gamma rays are useful in shrinking tumor and destroying cancerous cell in human body [5,7].

Any staffs if assigned to be worked in a radiological department should poses adequate knowledge on radiation safety. Continuous professional development programmes should be provided to refresh and to increase their knowledge and awareness on the safety of ionizing radiation from time to time. There are a few reported studies on the level of radiation knowledge and radiation safety awareness amongst medical staffs in other countries [1, 8, 9, 10]. However, the level of knowledge and radiation safety awareness among nuclear medicine nurses in Malaysia and the effectiveness of the training program are not known. The aims of this study were (1) to determine the level of knowledge and awareness on radiation safety among the Malaysian nuclear medicine nurses, and (2) to assess the effectiveness of a training program provided by hospital to increase the radiation knowledge and radiation safety awareness of the nuclear medicine nurses.

2. Materials and methods

The study took place at Putrajaya Hospital in Malaysia. A self-administered questionnaire was used in this study. The questionnaire was adapted from the previous studies [1, 11]. The questionnaire consists of two main areas, i.e., the general radiation knowledge (8 questions) and radiation safety (8 questions). Following the design of this questionnaire, the content was validated by medical physicists at the Nuclear Medicine Department of the hospital. The questionnaire was distributed to the nuclear medicine nurses during a one-day radiation safety training program organized by the hospital in November, 2013. The same set of questionnaire was administered twice, i.e. before and after the training program.

The completed questionnaires were assessed by one assigned researcher and scores were calculated on general radiation knowledge and radiation safety areas. Correct answers were awarded one mark, whereas an incorrect answer or omission received zero mark, hence resulting in a total score ranging from 0 to 16. Survey data were collected before and after the training and were analyzed using descriptive statistics and paired sample t-test. All the statistical analyses were performed using Statistical Package for Social Science (SPSS), version 21.0.

3. Results

3.1. Demographics

A total of 27 nurses comprising Trained Nurses and Community Nurses has responded to the questionnaire before and after attending the radiation safety training program. The general demographics of the participants for gender showed that 92.6% were female while 7.4% were male. 92.6% and 7.4% of the respondents were in the age range of 20 – 29 and 30 – 39 years old, respectively. In terms of academic qualifications, there were 14.8% of the respondents graduated with certificate level whereas 85.2% were diploma holders and 3.7% were degree holder. The respondents of this study were fresh graduates and newly joined the Ministry of Health of Malaysia. Therefore, this group of nuclear medicine nurses is identified to have very limited working experience with radiation (0 – 2 years).

3.2. Overall scores

The overall scores findings showed that the range of the total scores obtained by the nuclear medicine nurses before and after the training were 6-14 (mean score = 11.19, SD = 1.92) and 13-16 (mean score = 14.85, SD = 0.91) marks, respectively. It was found that there was a significant increase in terms of overall scores by the respondents after attending the training program ($p<0.001$). Based on the analysis of section scores, it was found that the respondents in general performed slightly better on the general radiation knowledge area as compared to radiation safety area. The mean scores obtained before training on general radiation knowledge area was 5.74 (SD = 1.35) and on the radiation safety area was 5.44 (SD = 1.28). The corresponding values achieved after training was 7.59 (SD = 0.57), and 7.26 (SD = 0.66), respectively. There was a significant increase in terms of participants' scores for both sections after training ($p<0.001$). With this, the researchers have enough evidence to conclude that the radiation safety training program was effective in raising the knowledge and awareness of the respondents on radiation safety.

3.3. Items addressing the level of general radiation knowledge of respondents
Table 1 shows the percentage of the correct answers of the respondents for each general radiation knowledge item before and after training. It was found that Item No. 7 is the easiest items, all the respondents are aware that unborn child is more sensitive to radiation as compared to other groups of people such as child, adolescent, and adult. Among the eight items listed in this section, it was noticed that all the respondents were able to answer correctly four items only after training, i.e., Item No. 2, 3, 4 and 7. It was interesting to find that Item No. 5 is the most difficult item in this section for the respondents. The percentages of the respondents that gave the accurate answer (no limit) before and after the training course were 14.8% and 74.1%, respectively. The distributions of the choice of answers marked by the respondents before and after training were shown in Figure 1 and Figure 2. Based on this result, the researchers would like to highlight that quite a number of our nuclear medicine nurses still unaware that there is actually no limit to the annual whole body dose received by patient although they have attended the training program.

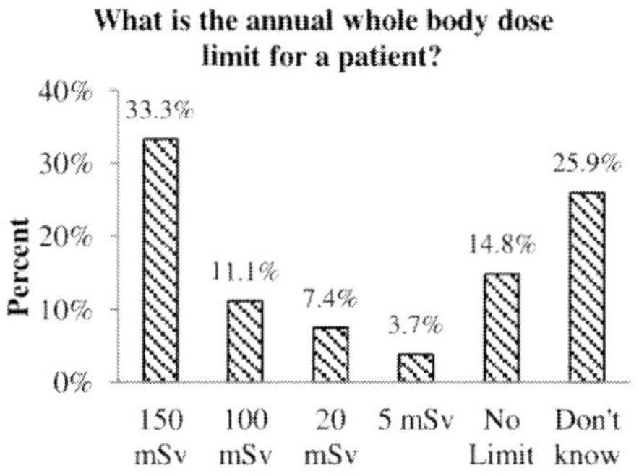

Figure 1. The distribution of the choice of answers for item "What is the annual whole body dose limit for a patient?" before training.

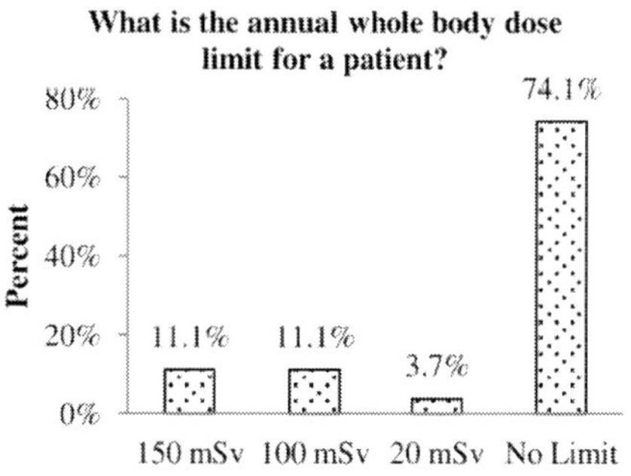

Figure 2. The distribution of the choice of answers for item "What is the annual whole body dose limit for a patient?" after training.

ALARA stands for "As Low as Reasonable Achievable". The ALARA principle was asked in Item No. 6. The percentages of the respondents that were able to give the correct answer for this item before and after the training were 85.2% and 88.9%, respectively. Although, the result showed that most of the nuclear medicine nurses are well aware of the principle, but it was found that the percentage of changed of the correct answer of the respondents for this item before and after training is not high (3.7%). This result is quite unexpected.

In the field of radiation, a dosimeter is a measuring device used to measure radiation dose, and it cannot be applied as a radiation protection tool. In item No. 8, respondents were asked to decide whether a personal dosimeter can provide protection from radiation exposure. It was found that not all the respondents are aware of it. The percentages of the respondents that gave the correct answer for this item before and after the training were 92.6% and 96.3%, respectively.

Table 1. Percentage of the correct answers of the respondents for general radiation knowledge before and after training.

Item No.	Item	%	
		Before training	After training
1	Radioactive radiation refers to Ionizing radiation	93	96
2	Which of the following radiations will be completely stopped by a piece of paper? Alpha particle	89	100
3	What is the unit of contamination? Bq/cm^2	40.7	100
4	What is the annual whole body dose limit for radiation worker? 20 mSv	63	100
5	What is the annual whole body dose limit for a patient? No Limit	14.8	74.1
6	Which of the following explains the ALARA principle? As Low as Reasonable Achievable	85.2	88.9
7	Which of the following is more sensitive to radiation? Unborn child	100	100
8	A radiation dosimeter provides protection from radiation exposure? False	92.6	96.3

3.4. Items addressing the level of radiation safety awareness of respondents

Table 2 shows the percentage of the correct answers of the respondents for each radiation safety awareness item before and after training. Among the eight items listed in this section, it was noticed that all the respondents were able to answer correctly five items only after training, i.e., Item No. 9, 12, 13, 14 and 15.

All respondents were able to give the correct dose limit for public in this item after the training compared to just 55.6% before attending the course. This showed that the respondents were well aware of the annual whole body dose limit for the public after the training. It was surprising to notice that the percentage of changed of the correct answer of the respondents for Item No. 14 before and after training is among the highest, i.e., 44.4%.

Based on the results, it was found that the two most difficult items in this section were Item No. 11 and Item No.16. This is because the percentages of the correct answers of the respondents for these two items after training were low.

If one increases the distance of a radiation source with its target, the amount of radiation exposure will decreases. Item No. 11 was designed to test the respondents' knowledge on this relationship. The percentage of the respondents that answered this item correctly also increased from 48.1% before the training to 77.8% after the training. It seems that there is still quite a number of respondents does not aware that this is also a way to protect them from radiation exposure.

There is a range of radiation safety items available in the hospital such as emergency tongs, radiation warning signs, survey meter, and safety manual. Among these four items, the respondents were asked in item No. 16 to choose one item that must be considered as the most important in the radiation safety. It is surprisingly to observe that only 51.9% of the respondents gave the correct answer (survey meter) to this question after the training compared to 40.7% before the training. This showed that the respondents have different point of view in terms of the priority of using a radiation safety item. The distributions of the choice of answers before and after training were shown in Figure 3 and Figure 4, respectively. Before training, it was observed that there were equal percentages (40.7%) of respondents chosen "radiation warning signs" and "survey meter" for this item, but none chosen emergency tongs. However, after the training, it was found that there was an increase in the percentage of respondents who chosen "survey meter" as the most important radiation safety items.

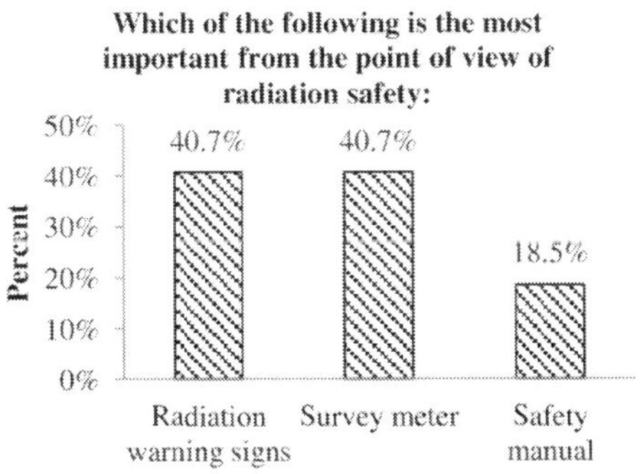

Figure 3. The distribution of the choice of answers for item "Which one is the most important for the point of view of radiation safety" before training.

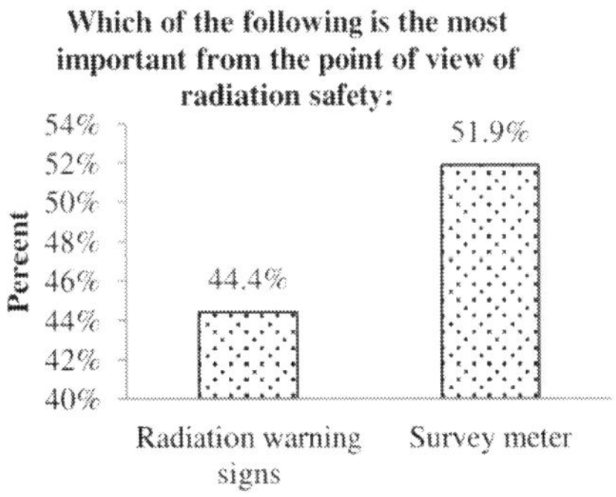

Figure 4. The distribution of the choice of answers for item "Which one is the most important for the point of view of radiation safety" after training.

Table 2. Percentage of the correct answers of the respondents for radiation safety awareness.

Item No.	Item	%	
		Before training	After training
9	Which of the following is the basic principles of radiation safety : Shielding, time and distance	77.8	9
10	Medical practitioner should .... the amount of the time in the room with radiation therapy patient in order to decrease radiation exposure. Decrease	48.1	10
11	As the distance between medical staff and radiation source increases, the radiation exposure .... Decreases	48.1	11
12	A shield between a gamma-emitting radiation source and medical staff causes the radiation exposure to .... Decrease	85.2	12
13	The radiation dose may be reduced by .... Moving further	92.6	13
14	What is the annual whole body dose limit for public? 1 mSv	55.6	14
15	Radiation Safety Office at respective department will be notified immediately if there is a radiation incident. True	96.3	15
16	Which one of the following items is most important from the point of view of radiation safety: Survey meter	40.7	16

4. Conclusion

From this survey, the training program provided by the hospital to the nurses in the nuclear medicine department was effective as there is a significant increase in the score of the respondents before and after the training. However, the knowledge and awareness on radiation safety among the nuclear medicine nurses are still at the moderate level.

It was further recommended that a national study be conducted to determine the level of knowledge about radiation and radiation safety awareness among all the medical staffs in Malaysia. Besides that future strategies to improve nuclear medicine nurses' knowledge and radiation safety awareness also need to be investigated, developed, implemented and evaluated.

Acknowledgement
The authors would like to acknowledge the co-operation of the staffs at the National Cancer Institute, Putrajaya. This work was supported by a grant from The Ministry of Higher Education (MOHE) under the Fundamental Research Grant Scheme (FRGS) 600-RMI/ST/FRGS5/3/Fst(218/2010).

References

[1] Eksioglu A S and Uner C 2012 Pediatricians' awareness of diagnostic of medical radiation effects and doses : are the latest efforts paying off? Diagnostic and interventional radiology **18** 78

[2] K E Thomas, E June, P Parmley, S Haidar, R Moineddin, E Charkot, G BenDavid, C Krajewski 2006 Assessment of radiation dose awareness among pediatricians *Pediatric radiology* **36** 823 – 832

[3] Mojiri M and Moghimbeigi A 2011 Awareness and attitude of radiographers towards radiation protection *Journal of Paramedical Science (JPS)* **2**(4) 2

[4] Wrixon A D 2008 New ICRP recommendations *Journal of Radiological Protection* **28** 161

[5] A Yucel, E Karakas, E Bulbul, I Kocar, B Duman and A Onur 2009 Knowledge about ionizing radiation and radiation protection among patients awaiting radiological examinations : A cross survey *The Medical Journal of Kocatepe* 25-31

[6] Ratnapalan S, Bona N, Chandra K and Koren G 2004 Physicians' perceptions of teratogenic risk associated with radiography and CT during early pregnancy *American Journal of Roentgenology* **182** 1107 – 1109

[7] Lawrence TS, Ten Haken RK and Giaccia A 2008 Principle of radiation oncology Cancer: Principles and Practice of Oncology. *8th ed. Philadelphia: Lippincott Williams and Wilkins*

[8] Shiralkar S, Rennie A, Snow M, Galland RB, Lewis MH and Gower-Thomas K 2003 Doctors' knowledge of radiation exposure : questionnaire study *BMJ* **327** 371 -372

[9] Soye JA and Paterson A 2008 A survey of awareness of radiation dose among health professionals in Northen Ireland *British journal of radiology* **81** 725 – 729

[10] Lee C I, Haims A H, Monico E P, Brink J A and Forman H P 2004 Diagnostic CT Scans : Assessment of Patient, Physician and Radiologist Awareness of Radiation Dose and Possible Risk 1 *Radiology* **231** 393 -398

[11] Quinn AD, Taylor CG, Sabharwal T and Sikdar T 1997 Radiation protection awareness in non-radiologists *Radiation protection* **70** 102 – 106

Measurement of terrestrial gamma radiation dose-rate (TGRD) level in soil samples from the district of Rembau, Malaysia, using high-purity Germanium detectors

N E Norbani[1], **N A Abdullah Salim**[2] **and A T Abdul Rahman**[1,3,4].

[1] Faculty of Applied Sciences, Universiti Teknologi MARA, 40450 Shah Alam, Malaysia
[2] Malaysian Nuclear Agency, 43000 Bangi, Malaysia
[3] Department of Physics, University of Surrey, GU27XH Guildford, United Kingdom
[4] Institute of Science, Universiti Teknologi MARA, 40450 Shah Alam, Malaysia

E-mail: ahmadtaufek@salam.uitm.edu.my

Abstract: This study assesses the gamma radiation levels and associated dose rates from the naturally occurring radionuclides ^{232}Th, ^{238}U and ^{40}K in soil samples collected from the district of Rembau, Malaysia using High-purity Germanium (HPGe) Detectors. A 105 measurement were performed on surface soil using NaI (Tl) gamma-ray detector with crystal size 1"x1", covering about 83% land of the Rembau district. The concentration of the naturally occurring radionuclides ^{232}Th, ^{238}U and ^{40}K in soil samples collected were determined by using HPGe detector based on high-resolution gamma spectrometry system at Malaysia Nuclear Agency. The range of natural gamma radiation measured was from 114±14 nGy h^{-1} to 857±14 nGy h^{-1}. The range of activity concentrations of ^{238}U, ^{232}Th and ^{40}K in soil from the studied areas varies from 151-401 Bq kg^{-1}, 113-342 Bq kg^{-1} and 674-1526 Bq kg^{-1} with mean values of 245 Bq kg^{-1}, 186 Bq kg^{-1} and 1152 Bq kg^{-1} respectively. The mean values of terrestrial gamma radiation dose rate measured in Rembau district is 383±18 nGy h^{-1} compared to the Malaysian average is 92 nGy h^{-1} and world average is 59 nGy h^{-1} (UNSCEAR, 2000). The average annual dose from such terrestrial gamma radiation dose rates to an individual in Rembau district, assuming a tropical rural setting is estimated to be 0.78 mSv per year, which is considered to be within the normal range for doses from natural sources. An isodose map for the Rembau district has been plotted.

1. Introduction

Knowledge of various radionuclides in soil and rocks plays an important role in health physics and geo-scientific research [1]. The naturally occurring radionuclides ^{238}U, ^{232}Th and ^{40}K are the main sources of radiation in soils and rocks from which traditional building materials are derived. These radionuclides cause exposure risks externally due to their gamma-ray emissions [2]. Even though these radionuclides are widely distributed, their concentrations have been available depend on the local geological conditions and as such they vary from place to place [3]. This study assesses the level of terrestrial gamma radiation dose rate (TGRD) and associated dose rates from the naturally occurring radionuclides ^{232}Th, ^{238}U and ^{40}K in soil samples collected from the district of Rembau, Malaysia using High-purity Germanium (HPGe) Detectors.

The study examines relationship between TGRD levels, soil types and underlying geological formations in the district of Rembau, Malaysia. Studies agree that specific levels of TGRD are related to the composition of each lithological areas separated, and to the content of the rock from which the soils originate [4]. Yet, there is still no complete data of TGRD for the district of Rembau, Malaysia. The information provided will be used as a base line data for future reference as well as to predict the external gamma-ray dose rates throughout Malaysia without resorting to the full scale, laborious measurement from the ground. There is no doubt that the rapid development of the city of Kuala Lumpur has contributed towards the development of a new town in the district of Rembau. Accordingly, the TGRD information is very important as reference data for the background radiation dose, especially after reclamation is done for the development. This is also important in case of radioactive contamination occurred in the district of Rembau.Rembau district is located between latitudes 2° 25' and 2° 43' North and longitude 102° 00' and 102° 13' East. It has an area approximately 440 km^2 and a population of about 41 000 persons [5]. Seventy per cent of the area is covered by forest and the main land use is for agriculture. However the development of new town throughout the district of Rembau has been started recently.

Rembau district can be divided into three major geological formations of different geological age [6] as shown in Figure 1 and Table 1. The geological formations overlaid are, (a) Devonian, (b) Silurian and (c) Acid Intrusive Rock. Devonian (G5) and Acid Intrusive Rock (G7) are most abundant in Rembau district, about 60 per cent of the area.

Table 1. Geological formations of Rembau district, Malaysia.

Geological Label	Geological Name	Composition
G5	Devonian	Phyllite, schist and slate, limestone and sandstone, locally prominent. Some interbeds of conglomerate, chert and rare volcanic.
G6	Silurian	Schist, phyllite, slate and limestone. Minor intercalations of sandstone and volcanic.
G7	Acid Intrusive	Undifferentiated of granite and intrusives rock.

Rembau district is overlaid by six group of soil types as classified by FAO/UNESCO [7] as shown in Figure 2 and Table 2. The soil types are;

(a) Nitosol: Nitosols is a soil of shiny pad surfaces. There are three types of Nitosols, namely Dystric Nitosols (Renggam), Rhodic Nitosols (Tavy) and Haplic Nitosols (Serdang). Dystric Nitosols is the most abundant soil type in Rembau district.

(b) Ferralsols: Ferralsols is a soil with high content of sesquioxides. There are two types found in the Rembau district. These soil types are characterized as Orthic Ferralsols and Plinthic Ferralsols. It is also called Munchong and Malacca.

(c) Acrisols: Acrisols is an acidic soil of low base saturation. Two types are found in the southern part of Rembau district. They are characterized, as Orthic Acrisols and Plinthic Acrisols, and the local names are Batu Anam and Durian respectively.

(d) Fluvisols: Fluvisols; this group consists of flood plains and alluvial soils. There is one type of Fluvisols, namely Thionic Fluvisols, and the local names is Kranji. Most of these groups are found on the coastal plain, mostly in tidal swamps covered by mangrove.

(e) Leptosols: This group consists of clay in all horizons to a depth of 100cm or more. There is only one type of this group in Rembau district. Dystric Leptosols or local name called Seremban Series.

(f) Miscellaneous soils, (i) Steep Land. (ii) Local Alluvium: Most of Rembau district is steep land with forested area above 30 m in height [8].

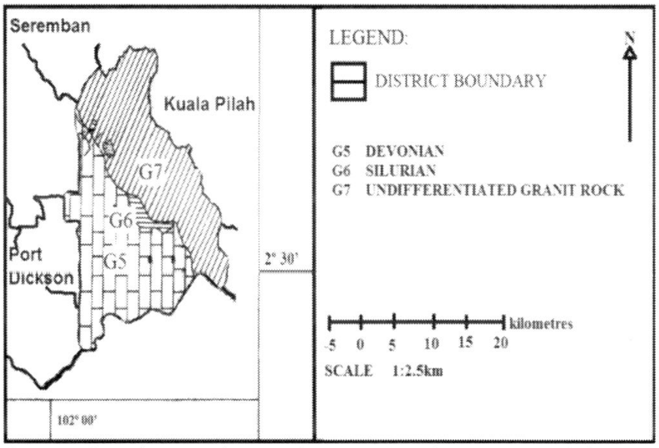

Figure 1. Map of geological features in Rembau district of Negeri Sembilan state, Malaysia [6].

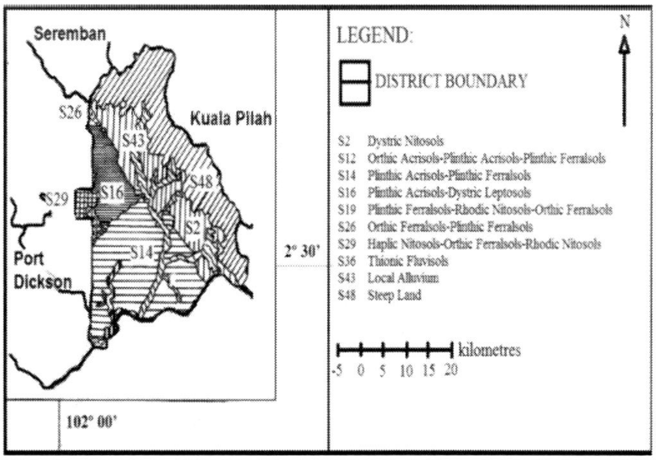

Figure 2. Map of soil types in Rembau district of Negeri Sembilan, Malaysia [7].

Table 2. Soil types of Rembau district, Malaysia.

Soil types label	FAO units	Local name
S2	Dystric Nitosols	Renggam
S12	Orthic Acrisols-Plinthic Acrisols-Plinthic Ferrasols	Batu Anam-Durian-Malacca
S14	Plinthic Acrisols-Plinthic Ferrasols	Durian-Malacca
S16	Plinthic Acrisols-Dystric Leptosols	Durian-Seremban
S19	Plinthic Ferrasols-Rhodic Nitosols-Orthic Ferrasols	Malacca-Tavy-Munchong
S26	Orthic Ferrasols-Plinthic Ferrasols	Munchong-Malacca
S29	Haplic Nitosols-Orthic Ferrasols-Rhodic Nitosols	Serdang-Munchong-Tavy
S36	Thionics Fluvisols	Kranji
S43	Local Alluvium	Local Alluvium
S48	Steep Land	Steep Land

2. Methods

Measurement of gamma radiation levels will made at each crossing point of the latitude and longitude lines. Rembau district will align the grid into squares of area around 1' x 1' (about 1.85 km x 1.85 km). The location for each sampling point will establish by global positioning system (GPS) (Garmin eTex HC series) [9] and survey gamma radiation levels at the field.

Measurements of the gamma radiation levels were conducted with gamma ray detectors manufactured by Ludlum of the U.S.A using model 19, Micro R Meter. The equipment uses a 2.54 x 2.54 cm NaI crystal doped with thallium [NaI(TI)], which responds to gamma radiation. The smallest scale division for the instrument is 1 μR h^{-1} (~ 9 nGy h^{-1}) [10]. The instrument had almost a flat energy response to gamma radiations between 40 keV to 1.2 MeV [11]. It is suitable for environmental gamma radiation measurements. It covers majority of significant gamma radiations emitted from terrestrial sources [12]. The uncertainty of reading observe on the maximum scale of the instrument is of the order of 10%. The measurements of TGRD survey is taken away from sites of developments such as road, building and foundation soils.

Six soils samples were collected covering the high, medium and low level of terrestrial gamma radiation dose-rate measured in situ. The soil samples were collected at a depth of 5 cm of surface soils. The samples were packed in a plastics bag, label with the date, location and bring in the laboratory for analysis. Move the stones, pebbles and organic materials from the sample then the samples will dry in an oven at about 100 °C for 24 hours to remove the moisture content and crush to pass through a fine mesh sieve (315 μm) to homogenize size. Then, it was deposited into a marinelli beaker and seal with a PVC tape and store about a month before counting. The marinelli beaker is use because of their ability to provide good detection sensitivity. Each sample will analyze; this is done as part of quality assurance and quality control of the analytical technique and as test of analytical precision [9]. ^{238}U, ^{232}Th and ^{40}K concentration will be measure using high-purity germanium detectors.

The activities of these samples and standards will be measured by using a horizontal hyper germanium detector coupled to an Analog to Digital Convertor (ADC) facilities at radiation laboratory, Agency Nuclear Malaysia. A Nuclear Data 66 computerised Multi-channel analyser will use for pulse height analysis. This counting system has an energy resolution of 1.00 keV adjust to the full width at half maximum (FWHM) of the 1332 keV gamma ray of ^{60}Co [9].

Count rates were recorded for seven gamma energies: 0.239 MeV (^{212}Pb, with a 44.6% gamma yield); 0.352 MeV (^{214}Pb, 37.1%); 0.609 MeV (^{214}Bi, 46.1%); 0.911 MeV (^{228}Ac, 27.7%); 1.461 MeV (^{40}K, 10.7%); 186 keV (^{226}Ra, 3.41%) and 583.1 keV (^{208}Tl, 86.0%). Concentrations of ^{232}Th were determined from the average concentrations of ^{228}Ac, ^{212}Pb and ^{208}Tl in the sample, and ^{238}U was determined from the average of the ^{226}Ra, ^{214}Bi and ^{214}Pb concentration.

Finally, using the WinSurf 32 (Golden Software Company), the isodose map of the terrestrial gamma radiation dose rate levels has been plotted for Rembau district.

3. Results

The distribution of the gamma radiation levels data is presented by histogram in Figure 3. The most frequently recorded value were at 300 and 600 nGy h^{-1}.

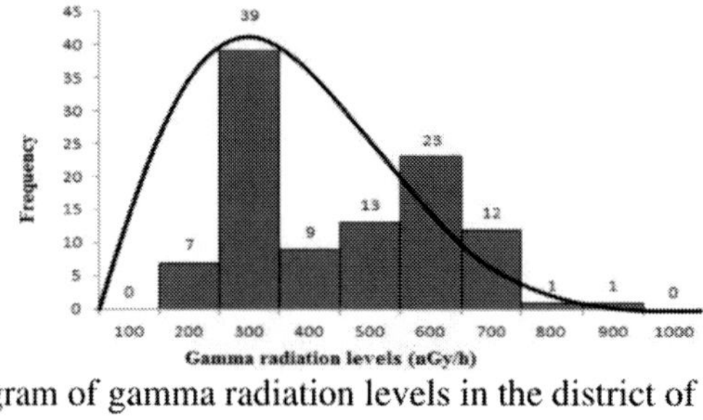

Figure 3. Frequency histogram of gamma radiation levels in the district of Rembau, Negeri Sembilan.

Table 3 and Table 4 show mean gamma radiation levels for each different geological features and for each different soil types. Acid Intrusives Rock (G7) gives the highest mean gamma radiation levels, which is 536 nGy h^{-1}. Dystric Nitosols (S2), Local Alluvium (S43) and Steep Land (S48) were the most abundant soil types covering of Rembau district. They give the highest mean gamma radiation levels measured that are 493 nGy h^{-1}, 506 nGy h^{-1} and 547 nGy h^{-1} respectively. The lower mean gamma radiation levels were found in the areas covered by soil types Haplic Nitosols-Orthic Ferrasols-Rhodic Nitosols (S29) and Plinthic Ferrasols-Rhodic Ferrasols (S19), that is 221 nGy h^{-1} and 229 nGy h^{-1} respectively. The area in Rembau district covered by Devonian geological features show the lowest mean gamma radiation levels measured, which is 274 nGy h^{-1}.

The highest gamma radiation levels measured in Rembau district is 857 ± 14 nGy h^{-1}. It was found in area covered by soil type Dystric Nitosols (S2) and Local Alluvium (S43) with underlying geological features Acid Intrusive (G7). These soil type of high gamma radiation levels is abundant in granitic areas, which are extensively intruded by schist, shale, quartzite and siltstone. Granites usually have U, Th and K contents higher than other crustal rocks [13]. Dystrict Nitosols or locally named Renggam also has 43% clay content and the highest K concentration in 0 – 8 cm depth from surface [8]. Both of these soil type are found in area with underlying of Acid Intrusive Rock (G7). Acid Intrusive Rock is rock that has crystallized from molten rock material called magma. These rocks contain higher monazite, basalt and rhyolite [12]. Therefore the highest gamma radiation levels were found in the area covered by those soil types and geological features. It was found in area covered by Haplic Nitosols-Orthic Ferrasols-Rhodic Nitosols (S29) with underlying geological features Devonian (G5).

Table 3. Mean gamma radiation levels measured for each different geological features covering the Rembau district, Malaysia.

Geologica l features	Gamma Radiation Levels (nGy h^{-1})					95% confidence interval for mean	
	Mean	Standard deviation	Minimum	Maximu m	Standar d error	Lower bound	Upper bound
G5	274	114	114	629	15	244	304
G6	400	149	300	571	86	30	770
G7	536	19	214	857	19	498	674

Table 4. Mean gamma radiation levels measured for each different soil types covering the Rembau district, Malaysia.

Soil types	Gamma Radiation Levels (nGy h^{-1})					95% confidence interval for mean	
	Mean	Standard deviation	Minimum	Maximum	Standard error	Lower bound	Upper bound
S2	493	106	257	686	23	446	540
S12	264	71	214	314	50	294	234
S14	245	82	157	629	14	216	274
S16	288	90	214	486	34	205	371
S19	229	40	200	257	29	199	259
S26	295	71	214	243	41	245	345
S29	221	214	114	543	107	181	261
S43	506	169	214	857	51	392	620
S48	547	130	214	714	30	484	610

The lowest gamma radiation levels measured in Rembau district is 114 ± 14 nGy h^{-1}. Lower gamma radiation levels were registered on Haplic Nitosols-Orthic Ferrasols-Rhodic Nitosols (S29) and Plinthic Ferrasols-Rhodic Nitosols-Orthic Ferrasols (S19) with underlying of geological features Devonian (G5) in certain location along the river. These soil types was overlaid by Devonian geological structure, which was formed from phyllite, schist and slate, limestone and sandstone, locally prominent as peat, humic clay and silt. Gamma radiation levels generally are lower in Devonian area. This is similar to the previous study at Melaka [14].

Following the spectrum analysis, the count rates for each detected photopeak and activity per mass unit (specific activity or radiological concentration) for each of the detected nuclides are calculated. The specific activity (in Bq kg^{-1}), A_{Ei}, of a nuclide I and for a peak at energy E, is given below:

$$A_{Ei} = \frac{N_{Ei}}{(M_E \times t \times Y_d \times M)}, \quad (1)$$

where N_{Ei} is Net Peak Area of a peak at energy E, M_E is the detection efficiency at energy E, t is the counting livetime, Y_d is the number of gammas per disintegration efficiency of this nuclide for a transition at energy E, and M_s is the mass in kg of the measured sample [10].

Table 5. The activities concentrations of the ^{238}U, ^{232}Th and ^{40}K in Bq kg^{-1} for the soil samples in Rembau district.

Sample No.	^{238}U [Bq kg^{-1}]	^{232}Th [Bq kg^{-1}]	Ratio U/Th	^{40}K [Bq kg^{-1}]
1	230	149	2:1	674
2	285	152	2:1	894
3	402	342	1:1	1136
4	188	150	1:1	1455
5	214	213	1:1	1526
6	151	113	1:1	1226
Mean	**245**	**186**	**1:1**	**1152**
World Average (15)	**35**	**30**	**3:1**	**400**

Table 5 shows the range of activity concentrations of ^{238}U, ^{232}Th and ^{40}K in soil from the studied areas varies from 151-402 Bq kg^{-1}, 113-342 Bq kg^{-1} and 674-1526 Bq kg^{-1} with overall mean values of 245 Bq kg^{-1}, 186 Bq kg^{-1} and 1152 Bq kg^{-1} respectively. The ratio of uranium to thorium is calculated to be 1:1 compared to the world average is 3:1[15]. Figure 4(a) shows the graph of uranium concentration vs. thorium concentration obtained in Rembau. Increase concentration of uranium, thorium and potassium in the crust is generally when the content of SiO_2 and K_2O [16]. It is important to assess the gamma radiation hazards to human associated with the used these samples for buildings; these were done by calculating the different radiation hazard indices.

The absorbed dose-rate, D_c (nGy h^{-1}), at the surface soil due to the concentrations of ^{238}U, ^{232}Th and ^{40}K in soil in all sampling locations is presented in Figure 4(b). The dose can be calculated using the absorbed dose-rate activity conversion factors depending on the radionuclides studies in the soil. The convesion factor by [15] was adopted and calculated using the equation given below:

$$D_c = 0.427\, A_U + 0.662\, A_{Th} + 0.043\, A_K \quad (2)$$

where A_U, A_{Th} and A_K are the activity concentrations of primordial radionuclide (^{238}U, ^{232}Th and ^{40}K) existing in the soil in Bq kg^{-1}. The straight line from Figure 4(b) show that the correlation coefficient indicates good general agreement between measured and calculated dose rates obtained.

The ultimate use of the measured activities in building materials is to estimate the radiation dose expected to be delivered externally if a building is constructed using these materials. To limit the annual external gamma-ray dose to 1.5 mSv y^{-1} [15], the external hazard index (H_{ex}) is given by the following equation:

$$H_{ex} = \frac{A_U}{740} + \frac{A_{Th}}{520} + \frac{A_K}{9620} \leq 1 \tag{3}$$

where A_U, A_{Th} and A_K are the concentrations of ^{238}U, ^{232}Th and ^{40}K, respectively, in Bq kg^{-1}. Radiation hazard are assume negligible if the value is less than unity.

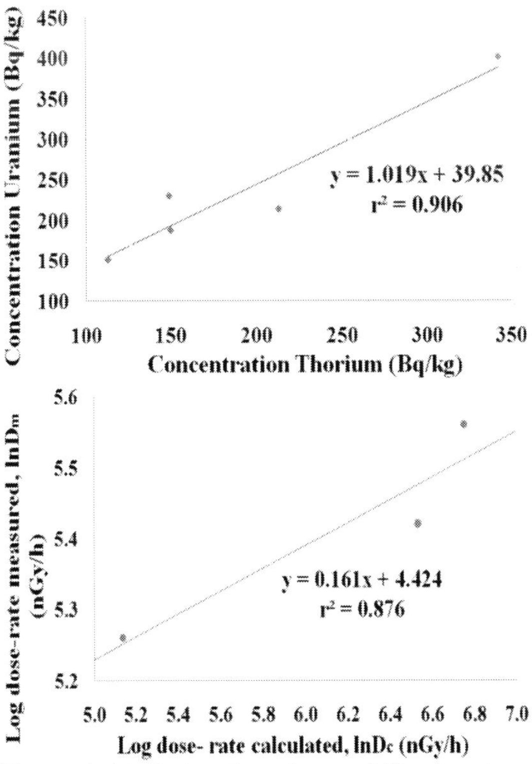

Figure 4. (a) Ratio of uranium and Thorium in soil sample of Rembau district and (b) Correlation graph of log dose-rate measured, $\ln D_m$ versus log dose-rate calculated, $\ln D_c$.

From Table 6 also recorded the external hazard index (H_{ex}) for the study area. The calculated values of H_{ex} for the soil samples in this studied are in range 0.54-1.32. The value of this index must be less than unity in order to keep the radiation hazard to be insignificant. Since the values of the study area are less than unity, therefore, according to Radiation Protection 112 report, this region is assume safe to population.

Outdoor gamma radiation levels measured throughout of Rembau district was conducted in this study. Figure 5 shows the isodose map of gamma radiation levels measured was drawn to present environmental gamma radiation levels distribution in Rembau district. A comprehensive understanding of spatial distribution of gamma dose rate is essential in assessing potential human risk associated with surface soil contamination by radionuclide. It is important for determining radiation detriment to the population as a whole.

Table 6. The measured values of dose rates (D_m) and external hazard index (H_{ex}) of the samples under investigation in Rembau district.

Sample No.	Soil	Geology	Measured dose-rate, D_m [nGy h^{-1}]	External Hazard Index (H_{ex})
1	S2	G7	686	0.66
2	S43	G7	857	0.77
3	S43	G7	486	1.32
4	S2	G7	486	0.69
5	S14	G5	200	0.85
6	S14	G5	171	0.54
	Mean		**383**	**0.80**

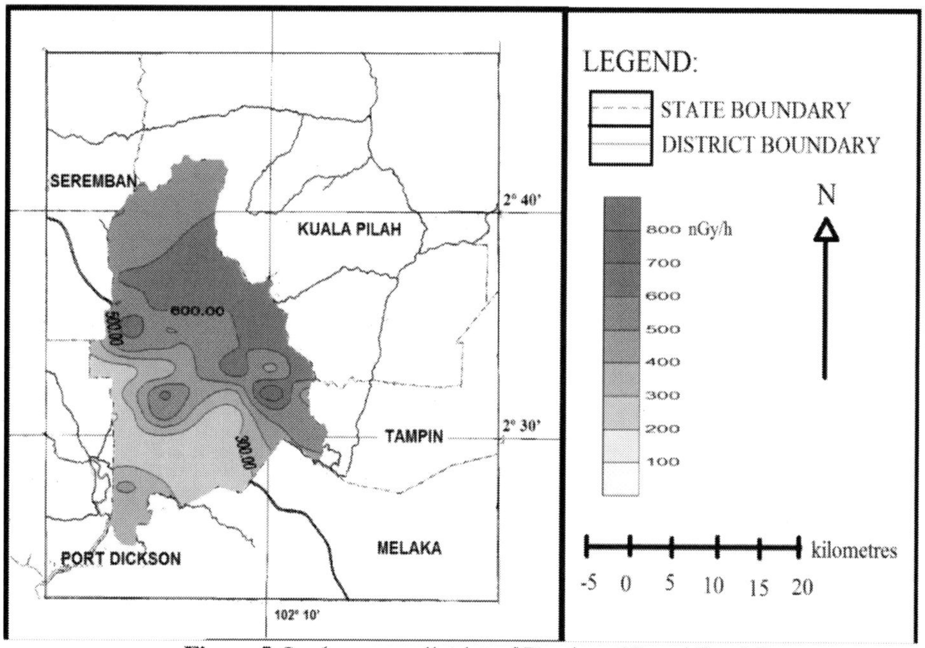

Figure 5. Isodose map district of Rembau, Negeri Sembilan.

4. Conclusions

The average terrestrial gamma radiation dose rates measured in Rembau district is 383 ± 18 nGy h^{-1} and the population weighted mean dose rate Rembau is 316 nGy h^{-1}. The Malaysian average is 92 nGy h^{-1} and world average is 59 nGy h^{-1} [15]. The finding show the average was 4 times higher than Malaysian average and 6 times higher than world average. The range of activity concentrations of ^{238}U, ^{232}Th and ^{40}K in soil from the studied areas varies from 151-402 Bq kg^{-1}, 113-342 Bq kg^{-1} and 674-1526 Bq kg^{-1} with overall mean values of 245 Bq kg^{-1}, 186 Bq kg^{-1} and 1152 Bq kg^{-1} respectively. The total absorbed dose rates in air calculated from the concentrations of the three radionuclides ranged from 192 nGy h^{-1} to 447 nGy h^{-1} and the external annual effective dose rate of the areas were determined to be between 0.39-0.91 mSv y^{-1}. The results can be considered as base values for distribution of natural radionuclides in the region and will be used as reference information to assess any changes in the radioactive background level due to geological processes. Using the conversion factor of 0.7 Sv/Gy [15] the average dose from such terrestrial gamma radiation dose rate to an individual assuming a tropical rural setting is estimated to be 0.78 mSv per year, which is considered to be within the normal range for doses from natural sources. It is not expected to cause statistical significance radiology health impact.

Acknowledgement
The authors would like to thank the Universiti Teknologi MARA, Malaysia (UiTM) for the Excellent Research Fund Scheme under Grant 600-RMU/ST/DANA 5/3/DST (12/2012) and the Ministry of Education Malaysia for supporting a research studentship for Nor Eliana Norbani under Mymaster [KPT (B) 890521115338]. A special gratitude goes to the State Government of Negeri Sembilan especially various government officials, and also to various owners of the lands, plantations and estates who have allowed us access to their properties to complete the ground survey.

References
[1] M Eisenbud *Environmental Radioactivity* New York : McGraw-Hill Book Company, 1963. pp. 135-163.
[2] UNSCEAR. *Sources and Effects of Ionizing Radiation* United Nations, New York. : United Nations ScientificCommittee on the Effects of Atomic Radiation, 1988.
[3] A C Tennisseen *Nature of earth materials* Printice-Hall, 1994. pp. 333-334. **NJ07632.**
[4] A T Abdul Rahman, M Z Jaafar, A T Ramli. M N Hasan. Kusadasi Turkey : *The International Nuclear Chemistry Congress 2005*; 22 May - 30 May 2005, 2005.
[5] Department Geological and Geographysics,2000.
[6] Director General of Geological Survey. *Map of Geological Features in Negeri Sembilan, Malaysia*. 1985.
[7] Director General of Agriculture Peninsular Malaysia. *Map of Soil Types in Negeri Sembilan, Malaysia*. 1973.
[8] Yaacob and Jusop. *Soil Science*. Kuala Lumpur : Dewan Bahasa and Pustaka, 1982.
[9] A T RAmli, A T Abdul Rahman, Abd Khalik Wood. Johor Bharu : s.n., 2003. The Proceeding of Annual Fundamental Science Seminar 2003; *Towards the Fundamental Research*. 20-21 May 2003. pp. 247-256.
[10] A T Ramli. Environmental Terrestrial Gamma Radiation Dose and its Relationship with Soil Type and Underlying Geological Formations in Pontian District, Malaysia. 1997, *Pergamon*, pp. 407--412.
[11] G K Knoll. *Radiation Detection and Measurement* (second edition). New York : Wiley and Sons, 1989.
[12] Radiological Characteristics of a Village on Uraniferoud Granitic Ground in Finland. A Voutileinen, O Castren, I Makelainen, K Winqvisk and H Arvela. 1988, *Radiation Protection Dosim.*, pp. 333-337.
[13] Zaini Hamzah, Ahmad Saat, Noor Hayati Mashuri and Seh Datul Redzuan. Surface radiation dose and radionuclide measurement inex-tin mining area, Kg Gajah, Perak. 2008, *The Malaysian Journal of Analytical Sciences*, pp. 419 - 431.
[14] A T Ramli, Sallehudin Sahrone and Husin Wagiran. Terrestrial gamma radiation dose study to determine the baseline for environmental radiological health practices in Melaka state, Malaysia. 2005, *Radiological Protection.*
[15] UNSCEAR. *United Nation Scientific Committee Of The Effect Atomic Radiation Report on The General Assembly*. New York : United Nation, 2000. pp. 1-112.
[16] Y Narayana, H M Somashekarappan, N Karunakara, D N Avadhani, H M Mahesh and K Siddhappa Natural Radioactivity In The Soils Sample Of Coastal Karnataka Of South India. 2001, *Health Physics*, pp. 24-33.
[17] Director of National Mapping. *Map of Negeri Sembilan, Malaysia* c type 2nd ed. Kuala Lumpur, Malaysia. 1989.

Journal of Physics: Conference Series **546** (2014) 012017 doi:10.1088/1742-6596/546/1/012017

Assessment of Ge-doped optical fibres subjected to x-ray irradiation

S A Ibrahim[1], S S Che Omar[1], S Hashim[1], G A Mahdiraji[2], D A Bradley[3, 4], A B Kadir[5] and N M Isa[5]

[1]Department of Physics, Universiti Teknologi Malaysia, 81310 Skudai, Johor, Malaysia
[2]Department of Electrical Engineering, Faculty of Engineering Building, University of Malaya, 50603 Kuala Lumpur, Malaysia
[3]Department of Physics, University of Surrey, Guildford, GU2 7XH, U.K
[4]Department of Physics, Universiti Malaya, 50603 Kuala Lumpur, Malaysia
[5]Medical Physics Group, Malaysian Nuclear Agency, Bangi 43000, Kajang, Selangor, Radiation Safety and Health Div, Malaysia

E-mail: saishahibrahim90@gmail.com

Abstract. We have reported the thermoluminescence (TL) response of five different diameters ~120, 241, 362, 483, and 604 µm of 6 mol percent Ge-doped optical fibres. The perfomance of the Ge-doped optical fibre are compared with commercially available TLD-100 chips (LiF:Mg,Ti) in terms of their sensitivity and minimum detectable dose (MDD). The irradiation was performed using X-ray machine (Model ISO 'Narrow Spectrum Series') provided by the Malaysian Nuclear Agency (MNA) at 60 kV X-ray irradiation in low doses ranging from 1-10 mGy. The results show the linear TL dose response from the fibres up to 10 mGy. The smallest diameter of 120 µm optical fibre shows the highest TL dose response compared to above mentioned fibres. The minimum detectable dose (MDD) is 0.82, 0.20, 0.14, 0.08, and 0.13 mGy for Ge-doped with diameters of 120, 241, 362, 483 and 604 µm. All TL materials show the MDD value within the delivered dose 0.01-1.00 mGy subjected to x-ray irradiation. The Ge-doped fibre with diameter of 483 µm was matched the MDD value of TLD-100 chips that equivalent to 0.08 mGy at the same irradiation. We have observed that among the five different diameters of optical fibre, 120 µm shows the best results and its better response than TLD-100 chips (by a factor of 5). The linear response at low dose levels makes this optical fibre most suitable for medical application.

1. Introduction

Thermoluminescence (TL) is well established technique of radiation dosimetry that allows the measurement of radiation dose to extremely low dose levels that are used in a various types of applications; medical, personal, environmental and industry. The determination of exposure of doses delivered to personnel working in the field of ionizing radiation fields is called Thermoluminescence Dosimetry (TLD). TLD is a basic for radiation protection which is to know the exact amount of absorbed dose delivered for optimization. Therefore, the study of new TL materials is important to improve the precision of the radiation measurement in specific applications.

The selection of the TL material type to be used in any application depends on the specific requirements of application. For example, improper use of TL materials in medical applications may lead to unacceptable bias and large uncertainties in the dose estimation [2]. Recently, a number of studies were investigated in Ge-doped optical fibre as new TL materials that have good linear response. It has been established to have potential in radiation dosimetry [3, 4]. To provide a good potential of TL material should be not only considered on radiation parameter such as energy, dose rate and dose response but also the type of fibres [5]. Up to now, optical fibres were studied to measure dose response for radiation dosimetry, in order to overcome spatial resolution that limited to a capability of ~ few mm existing in-vivo dosimetry system [3]. However, TL performance also influence by additional dopant to TL material which can greatly enhance the sensitivity of optical fibres to radiation by providing an increased number of traps. In additions, new defects and absorption bands can appear [4]. Besides, one advantage of optical fibre dosimeters is that they are impervious to water and in some instances, it then becomes possible to locate the optical fibre within a particular tissue of interest [6]. Measurement of the absorbed dose in tissue exposed to radiation, requires that the TL material has energy absorption characteristics nearly same as that of human tissue, i.e. to be tissue-equivalent. Nowadays, very few TL material are nearly tissue equivalent, however LiF are the famous tissue-equivalent detector that it equal to $Z_{eff} = 7.4$ for soft tissue. In this study was determined dosimetric characteristics for lithium borate, $Li_2B_4O_7$:Cu, in which obtained the effective atomic number of 7.8 which is closest energy absorption characteristics to tissue of all investigated TLDs in this work [7]. According to Furetta, the main dosimetric TL characteristics include glow curve, annealing procedure, TL sensitivity, dose response, precision of TL measurements, reproducibility, energy response, fading characteristics and minimum detectable dose. Its shows TLD are nearly to tissue equivalence to photon radiation, high sensitivity, very simple glow curve with a single well defined peak, low fading, excellent reproducibility and very wide linearity range [8].

In fact, Modified Chemical Vapor Deposition (MCVD) is the commonly method that is used to fabricate optical fibre began at Bell Laboratories. This process became rather widely practiced because it is simple, requires a minimum of equipment, and produces quality fibres. Subsequently, this process was developed to create new fibres design possibilities. MCVD process has an ability to precisely control index and spacing within the fiber core provided for efficient manufacture and so the process exists in spite of its limitations in deposition rate and total quantity of silica deposited. This is because of better control over refractive index profile and superior process flexibility [9].

In this paper, we present the experimental results to determine the TL sensitivity and minimum detectable dose (MDD) 6 mole percent of Ge-doped optical fibre with five different fibre diameters subjected to X-ray spectrum at low doses range. The purpose of this study is to investigate the suitability of optical fibres as a new TLD produced by X-ray irradiation. This results will be compared with standard dosimetry as a radiation dosimeter (LiF:Mg,Ti).

2. Materials and Methods

2.1 Sample preparation

The studied optical fibres and its original preform have been produced by the manufacturer through the MCVD process at University of Malaya (UM). The aim of this process is to design and produce tailored fibres, characterized in terms of radiation response and dependence on production parameters also examining the production possibilities. These allow for tailor-made fibres, producing enhanced TL yield per unit dose [10]. In this study, five different diameters of 6 mole percent Ge-doped optical fibres were used to comparing the TL dose response. The five diameters are ~120, 241, 362, 483, and 604 μm will be taken in this experiment. For comparison, the TLD-100 chips form is used in this research that has the dimension of 3.2 mm x 3.2 mm x 0.9 mm. In preparation the samples, optical fibre was cleaned using a cotton cloth dampened with small amount of methyl alcohol and the fibre was cut into 5.0 ± 0.1 mm long pieces using optical fibre cleaver (Fujikura, Japan). Then, the mass of fibres were recorded using an analytical balance BSA224S-CW (Germany) that are approximately 0.057, 0.625, 0.933, 2.3 and 2.7 ± 0.05 mg for ~120, 241, 362, 483, and 604 μm respectively. While,

23.76 ± 0.05 mg is the mass for TLD-100 chips. Vacuum tweezers were used for easy handling and grouping of fibres [3].

2.2 Annealing
Before exposure the 6 mole percent of Ge-doped optical fibres to the X-ray radiation they were placed in alumina crucible which was kept in an annealing oven. The fibres were annealed at 300°C for a period of 1 hour in order to erase all possible remaining information. During an annealing, about 24 hours is taken for cooling down inside the oven to avoid thermal crashes. The fibres were equilibrated at room temperature in the end of annealing process [11]. While, for TLD-100 chips, annealing were placed on a annealing plate for 1 hour at 400 °C and 2 hours at 100 °C. After annealing, the samples were storage inside dark plastic container to minimize environmental exposure such as humidity, temperature and visible light.

2.3 Irradiation
The materials were irradiated at 60 kV X-ray Tube that mostly used in medical diagnostic application. The irradiation was performed based on ISO 4037 'Narrow Spectrum Series' provided by Malaysian Nuclear Agency (MNA) with dose range 1 – 10 mGy. The optical fibres were placed on the surface of Solid Water phantom^{TM}. The phantom with outer dimensions 30 cm x 30 cm x 15 cm with PMMA walls (front wall 2.5 mm thick, other walls 10 mm thick) filled with water and known as the ISO water slab phantom. The fibres were positioned at the centre of radiation field and perpendicular to the beam axis with a fixed voltage at 60 kV. The slab phantom has a rectangular box shape with 15 cm x 15 cm and positioned at irradiation distance of 2 m. Filters according to ISO 4037–1 were used to produce the spectra of the radiation qualities of the ISO narrow spectrum series which were used 0.24 mm Cu and additional filter of 4.0 mm Al and 0.60 mm Cu. Hence, the total thickness of filtration used is 4.84 mm [12]. The dose rate of 12.444 mGy s^{-1} were using for delivered the dose. The fibres were handled carefully and were encapsulated in the gelatin capsules.

2.3 TL measurements
Post irradiation, the fibres were left for 24 hours to control thermal fading before readout. TL measurements were readout being performed using instrument Harshaw 3500 TL reader (Ohio, USA) and WinREMS software for displaying TL glow curve and TL response values at Malaysian Nuclear Agency (MNA). In this study, the time temperature profile (TTP) was set at preheat temperature of 50 °C for 8 s; acquisition temperature 300 °C for 13.33 s; heating rate cycle of 25 °C per second. Finally, an annealing temperature of 300 °C for 10 s was applied to sweep out any residual signal. All TL measurements were made in a nitrogen atmosphere in order to reduce the thermal noise resulting from the heating planchet of the TL reader. The TL yield obtained was then normalized to unit mass of the particular TL medium.

3. Results and Discussions

3.1 TL Dose response and Sensitivity
The dose response of fibre was analyzed in a fixed energy at 60 kV and different diameters of optical fibres. Figure 1 shows that the TL responses of optical fibre with five different diameters for low doses in the range of 1 to 10 mGy were exposed to X-ray radiation. All samples display a linear TL response for low doses in the range of 1 to 10 mGy. Assuming a linear fit, the change in TL yield per unit absorbed dose (in other words the slope) is 3.03, 1.51, 1.39, 1.34 and 1.31 TL.$mg^{-1}Gy^{-1}$ for 120, 241, 362, 483, and 604 μm respectively as shown in figure 1.

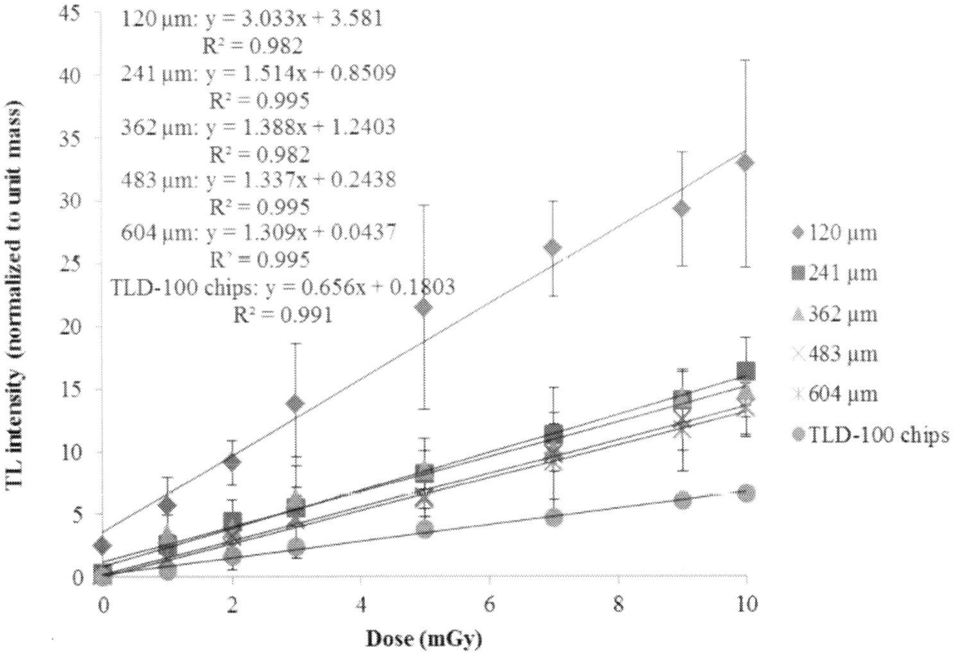

Figure 1 TL dose response of five different diameters optical

From the Figure 1, TL measurement presents a linear response over the dose range studied; however, for smallest diameter is slightly highest dose response and most sensitive followed by increasing diameters values. The variations values of these five different diameters were compared and clearly seen that they are approximately close to each other except for smallest diameter which has large gap than other fibres. The error bars represents the standard error of the mean, which for a fibre with 120 µm diameter is found to be in the range of 1.798-8.254. The finding indicates that at 3, 5 and 10 mGy the error bar was too large because it no longer precise. To conclude, the TL dose response increases with the smaller diameter. According to Bradley [12], the fibres performance of an irradiated optical fibre is influenced by the physical characteristic of the fibres include their spatial resolution of ~120µm, their water impervious nature giving the potential on interstitial and intercavitary measurement and the production of fibre through MCVD process that introduces dopant at selected concentrations.

In order to validating the results, the TL dose responses presented were compared with the TL response of TLD-100 chips as shown in Figure 1. As clearly seen that the all the fibres present higher response compared to TLD-100 chips which have TL response 0.66 $TL.mg^{-1}Gy^{-1}$. The dose response of 120 µm diameter show high response than TLD-100 chips by a factor of 5. This shows that the optical fibres are more preferable than TLD-100 chips in measurement of low dose. The implications were performed that the same irradiation of the 6 mole percent of GeO_2 optical fibre also influenced by parameter of radiation at lower energies. These X-ray measurements at least perform improvement to provide better and accurate dosimetry suitable for radiation protection of workers.

3.2 Sensitivity

In Figure 2 present the five different optical fibres of their sensitivity that were compared with TLD-100 chips in terms of $TL.Gy^{-1}.mg^{-1}$.

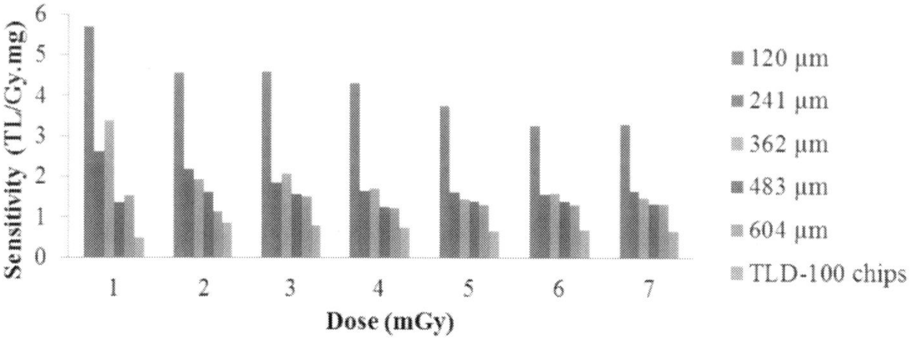

Figure 2 The sensitivities of different diameter sizes of optical fibres with TLD-100 chips.

In terms of sensitivity, the results illustriate a superior sensitivity of 120 μm than higher diameters fibres (TL dose response per dose). These results are also better sensitivities than TLD-100 chips. The relative sensitivities were determined to be 11.60, 5.36, 5.75, 5.70, 5.61, 4.83, and 5.00 for 120, 241, 362, 483, and 604 μm respectively with respect to TLD-100 chips. High spatial resolution diameter size of 120 μm was observed to be the highest sensitivity. The difference in the sensitivity at the higher dose is due to the low optical signal at these doses, giving a high signal-to-noise ratio. To conclude, 120 μm is suitable used for medical diagnostic application as the dose applied.

3.3 Minimum Detectable Dose

$$D_o = (B\ mean + 2\sigma)\ F \quad (1)$$

$$F = 1/m \quad (2)$$

Minimum Detectable Dose (MDD) is the smallest dose in response statically significant different from a background signal. Here, MDD were calculated using equation 1 which B_{mean} is the mean of TL signal obtained from the samples annealed but then not irradiated in mGy, σ is the standard deviation of the mean background. Meanwhile, *F* is the TL system calibration factor for each type of the optical fibres expressed in $GynC^{-1}$, which *m* were the slope obtained from TL dose response for each TL materials shown in equation 2 [14]. In the present study, MDD value was found to be 0.82 mGy, 0.20 mGy, 0.14 mGy, 0.08 mGy, 0.13 mGy, and 0.08 mGy for 120, 241, 362, 483, 604 μm, and TLD-100 chips respectively.

The MDD for all the samples include TLD-100 chips shows that it still quite well and far below 1 mGy (lowest delivered dose) using threshold dose. The Ge-doped fibre with diameter of 483 μm was matched the MDD value of TLD-100 chips that equivalent to 0.08 mGy at the same irradiation.. These values show a good detection of fibre at low doses range in X-ray. During the experimental of MDD, it is important the relation among high TL sensitivity, TL dosimeter material and the type of TLD reader used for obtaining a low detection dose.

4. Conclusions

The results of 6 mole percent of Ge-doped optical fibres with five different diameters are very promising for dosimetry applications. The Ge-doped optical fibres and TLD-100 chips exhibit good dose response up to 10 mGy irradiation. The sensitivity of five different diameters has been observed to be slightly higher than TLD-100 chips. Among the five different optical fibres, the fibre with 483 μm core diameter and TLD-100 chips have the lowest MDD i.e. 0.08 mGy. In future, our group has great intention to investigate the performance of Ge-doped optical fibres with different Ge concentration.

Acknowledgements
The authors are very much thankful to The Ministry of Education (MOE) HIR grant (A000007-50001) for funding the fiber pulling project and UTM for providing financial assistance through Research University Grant Scheme (RUGS), Project number (Q.J130000.2526.03H27).

References

[1] Kortov V 2007 *Radiat. Meas.* **42** 576-581.
[2] Moscovitch M and Horowitz Y S 2007 *Radiat. Meas.* **41** S71-77.
[3] Hashim S, Al-Ahbabi S, Bradley D A, Webb M, Jeynes C, Ramli A T, Wagiran H 2009 *Appl. Radiat. Isot* **67** 423-427.
[4] Yaakob N H, Wagiran H, Hossain Md I, Ramli A T, Bradley D A, and Ali H 2011 *Appl. Radiat. Isot.* **69** 1189–1192.
[5] Bradley D A, Abdullah Y A, and Amin Y M 2001 *Radiat. Phys. and Chem.* **61** 409-410.
[6] Yusoff A L, Hugtenburg R P, and Bradley D A 2005 *Radiat. Phys. and Chem.* **74** 459-481.
[7] Miljanic S, Ranogajec-Komor M, Knezevic Z and Vekic B 2002 *Radiat. Prot. Dosim.* **100** 437-442.
[8] Furetta C, Prokic M, Salamon R, and Kitis G 2000 *Appl. Radiat. Isot.* **50** 243-250.
[9] MacChesney M 2000 *IEEE Journal of Quantum Electronics* **6** 1305-1306.
[10] Bradley D A, Abdul Sania S F, Alalawia A I, Jafaria S M, M. Noor N, Hairul Azhar A R, Mahdiraji G A, Tamchek N, Ghosh S, Paul M C, Alzimam K S, Nisbet A, Maah M J 2014 *Radiat. Phys. and Chem.* In Press, Corrected Proof.
[11] Hashim S, Bradley D A, Saripan M I, Ramli A T, and Wagiran H 2010 *Appl. Radiat. Isot.* **68** 700–703.
[12] ISO 4037–1 International Standard 1997 Part 1: Radiation Characteristics and Production Methods.
[13] Bradley D A, Hugtenburg R P, Nisbet A, Abdul Rahman A T, Fatma Issa, Mohd Noor N, and Alalawi A 2012 *Appl. Radiat. Isot.* **71** 2-11.
[14] Khan F M, 2003. *The Physics of Radiation Therapy* (SA: Lippincott Williams & Wilkins).

Journal of Physics: Conference Series **546** (2014) 012018 doi:10.1088/1742-6596/546/1/012018

Thermoluminescent response of single mode optical fibre to x-ray irradiation

S S Che Omar[1], S Hashim[1], S A Ibrahim[1], W M S Wan Hassan[1], G A Mahdiraji[2], N Md Isa[3], M J Mad Isa[3], M M Abd Jalil[3] and A B Kadir[4]

[1] Department of Physics, Universiti Teknologi Malaysia, 81310 Skudai, Johor Darul Takzim, Malaysia
[2] Department of Electrical Engineering, University of Malaya, 50603 Kuala Lumpur, Malaysia
[3] Department of Medical Physics, Malaysian Nuclear Agency Bangi, 43000 Kajang, Selangor, Malaysia
[4] Secondary Standard Dosimetry Laboratory, Malaysian Nuclear Agency Bangi, 43000 Kajang, Selangor, Malaysia

E-mail: sitishuwaibah@yahoo.com

Abstract. We present the characteristics of the thermoluminescence (TL) response of single mode optical fibre (SMF) subjected to 30 and 70 kV x-ray irradiation. The TL responses are compared with commercially available TLD-100 (rod types). The SMF and TLD-100 were irradiated with x-ray source by using X-rays Generator model Phillips MG 165 located at Malaysian Nuclear Agency. The SMF and TLD-100 show linear dose response subjected to 30 and 70 kV x-ray irradiation. The SMF shows TL response by 10 times and 8 times greater than TLD-100 for the above-mentioned energies. The TL sensitivity characteristics of SMF show promising results to be introduced as a TL dosimeter material. The SMF could be used in several applications in the fields of medicine, industry, and research purposes.

1. Introduction

Thermoluminescence (TL) can be defined as the emission of light from matter following the previous absorption from the radiation source of exciting light energy, x-rays, or any other ionizing radiation. As being reported by Rivera (2012), the importance of TL for radiation dosimetry is due to the fact that the amount of light emitted is proportional to the absorbed dose by the irradiated material, which requires sensitive detection and accurate measurements of ionizing radiation [1]. Many researchers have investigated the performance of commercially available optical fibres subjected to photon [2-4] and x-rays [5-8]. In this study, the TL materials used is a tailor-made optical fibre which being fabricated using Modified Chemical Vapour Deposition (MCVD). Single mode optical fibre (SMF) (unknown dopant) has a circular symmetry and was fabricated mainly from pure silica (SiO_2). The basic construction for SMF is illustrated in Figure 1.

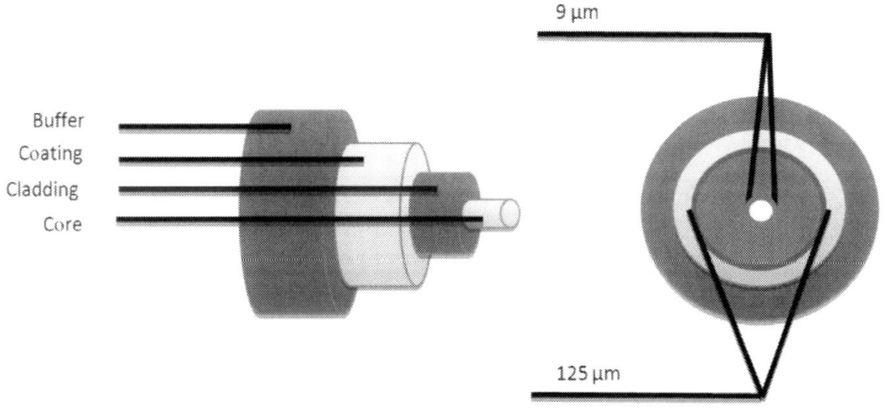

Figure 1. The structure of a typical single mode optical fibre.

The SMF usually has a core diameter between 8 to 10 µm and the cladding diameter is about 125 µm. These SMF have several advantages compared to commercial TLD-100, for instance, the fibres show high spatial resolution TL dosimetry (~ 100 - 120 µm), impervious to water (hygroscopic nature), flexibility and cheap relatively to be used in radiation dosimetry [3, 6].

According to The International Organization for Standardization, ISO (1999), it is recommended to use X-rays energies between 30 to 70 kV to treat tumours for diagnostics and industrial dosimetry purposes [9]. In this study, we have investigated the TL response of SMF at different x-ray energies (30 and 70 kV) and compare the TL response with the commercially available LiF:Mg, Ti (TLD-100).

2. Methodology

2.1. Material and preparation

Prior irradiation, a fibre stripper was used to remove the outer polymer coating of SMF. After removing the outer cladding, moist cotton was dipped into ethanol to clean the SMF and to avoid any presence of remnant polymer cladding. Then, the SMF was cut into approximate 0.5 ± 0.1 cm long pieces using optical fibre cleaver (Fujikura Ltd., Japan). To obtain the TL yield to be normalized per unit of mass, each of SMF (~ 0.25 ± 0.01 mg) and TLD-100 (rod) (~ 12.03 ± 0.01 mg) was measured using electronic balance (PAG Switzerland). The optical fibres were handling and grouping of the TL materials using vacuum tweezers (Dymax 5 – Charles Austen pump Ltd).

Annealing is the thermal treatment used to erase any irradiation history from the TL material, stabilizing the trap structure and restoring the dosimeter to initial conditions prior to irradiation. The annealing was performed with the optical fibres being kept in an oven set at 400 °C for a period of 1 hour, the optical fibres being retained in an alumina crucible during this time. Following the annealing cycle was the natural cooling down in the oven for 24 hours to avoid thermal stress; finally the optical fibres were equilibrated at room temperature. For the TLD-100, the annealing routine was to place these in a stainless steel plate and to anneal for 1 hour at 400 °C and subsequently 2 hours at 100 °C. After cooling, the TL materials were placed inside capsules and held within a light tight box in order to minimize exposure to potential high ambient light levels. Each capsule having at least five pieces fibre and the fibre stick on the wall of the capsule to make sure the fibre was irradiated uniformly.

2.2. Irradiation

In this study, after annealing, the sample of SMF and TLD-100 were exposed to the different energy of x-ray, which are 30 and 70 kV with a dose rate of 0.1869 mGy per second. As mentioned above, this study use x-ray energy of 30 and 70 kV to treat tumours for diagnostics and protection dosimetry purposes, respectively. The sample that irradiated to the both energies was obtained from the X-rays Generator model Phillips MG 165 placed at Malaysia Nuclear Agency (MNA), Bangi, Selangor. The X-ray machine has variable currents level up to 45 mA and a maximum voltage of 160 kV.

The readout process was performed at Department of Medical Physics, Malaysian Nuclear Agency (MNA). The TL measurements were carried out using a Harshaw 3500 TL reader (USA) with following parameters: time temperature profile (TTP) was set at preheat temperature of 50 °C for 10s; heating rate at 25 °C s^{-1} and acquisition time of 13.33 s. The maximum temperature that allowed in this reader is 300 °C remained for 20 s. The readings were taken under a gas flow of nitrogen. The nitrogen atmosphere was used to suppress possible spurious light signals from triboluminescence and also to reduce oxidation of the heating element.

3. Results and discussions

3.1. Glow curve

TL glow curve is defined as the plot of the amount of light emitted as a function of temperature [10]. In order to know the TL response of the SMF, this material was compared with the TL glow curve of TLD-100 (rod), using the same setup and read-out conditions. The comparison of glow curves is shown in Figure 2 and Figure 3. Figure 2 and Figure 3 illustrate the glow curve for energy of (a) 30 kV and (b) 70 kV for SMF and TLD-100 (rod) respectively.

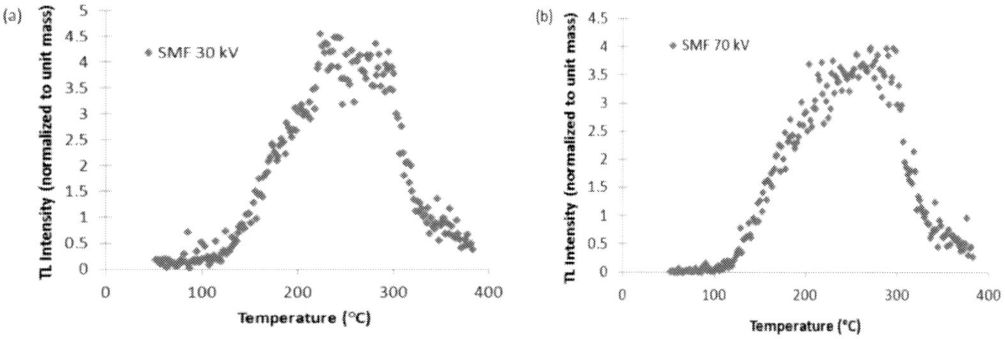

Figure 2. TL glow curves for SMF under (a) 30 kV and (b) 70 kV x-ray irradiations.

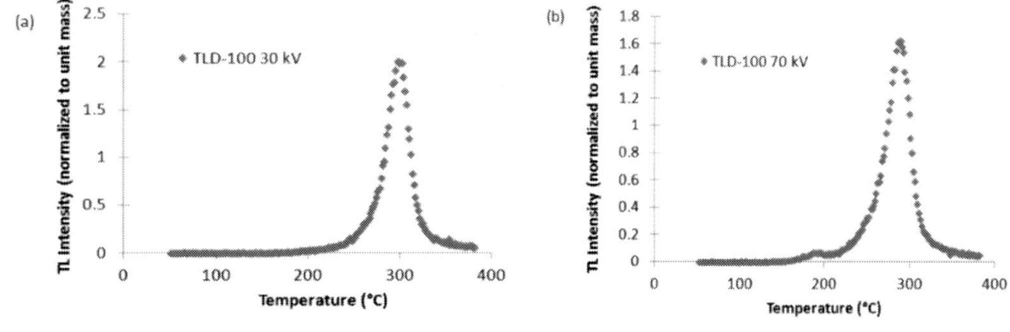

Figure 3. TL glow curves for TLD-100 (rod) under (a) 30 kV and (b) 70 kV x-ray irradiations.

The obtained results show a well-defined TL glow curve for SMF and TLD-100. In terms of shape, the SMF have a broad peak that present the characteristic of the amorphous media, while TLD-100 show a narrow shape. The shape (area under the curve) of both sample represent the energy absorbs after irradiated whereas the highest temperature peak is used for calculations. The TL glow peak for energy of 30 kV was obtained at 222 °C for SMF and 297 °C for TLD-100. Meanwhile for energy of 70 kV, the SMF and TLD-100 was observed at 287 °C and 289 °C with dose of 10 mGy, respectively. The general structure of TL glow curve remains unchanged by repeating the cycles of annealing and irradiation at various doses. As shown in figure 2, the intensity recorded for SMF 30 kV is 1.010 nA show greater intensity compare to SMF 70 kV (0.921 nA). According to Portal (1981), the TLD-100 have five typical peaks at 65 °C, 120 °C, 160 °C, 195 °C and 210 °C [11].

3.2. TL sensitivity

In this study, the TL sensitivity is expressed as TL dose response per unit of mass of dosimeter and per unit of x- ray doses (TL. $mg^{-1}.Gy^{-1}$). Figures 4 and Figure 5 show the TL dose response of SMF subjected to 30 and 70 kV x-ray irradiations. The cross-comparison was made with TLD-100 for radiation doses up to 10 mGy.

In the first part of the experiment, the irradiation was performed at 30 kV for SMF and TLD-100 (rod) as shown in Figure 4.

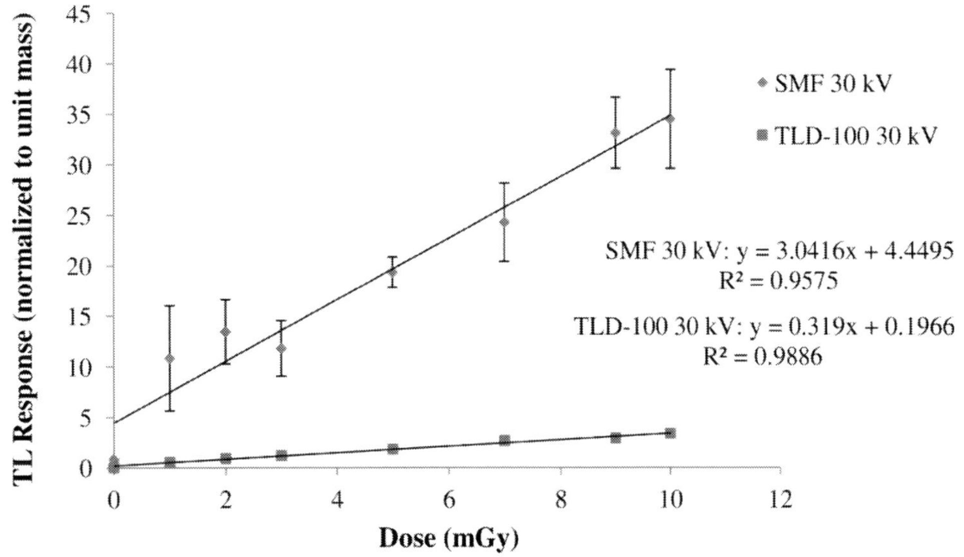

Figure 4. TL response against dose for SMF and TLD-100 (Rod).

The TL sensitivity for SMF was 10 times greater than TLD-100 (rod) i.e. the slope were 3.0416 $mg^{-1}.Gy^{-1}$ and 0.3190 $mg^{-1}.Gy^{-1}$ for SMF and TLD-100 (rod) respectively. The linear coefficient, R^2 was found to be 0.9575 for SMF and 0.9886 for TLD-100. Each single dose was corresponding to the mean of the three reading. The error bars on the graph indicate values to one standard deviation, finding by calculation. The error bar for SMF and TLD-100 is in the range of 0.049-5.185 and 0.004-0.354, respectively. It means the error bar for TLD-100 was too small to be seen in the graph.

In the second series of the experiment, the same TL materials were exposed to 70 kV as shown in Figure 5.

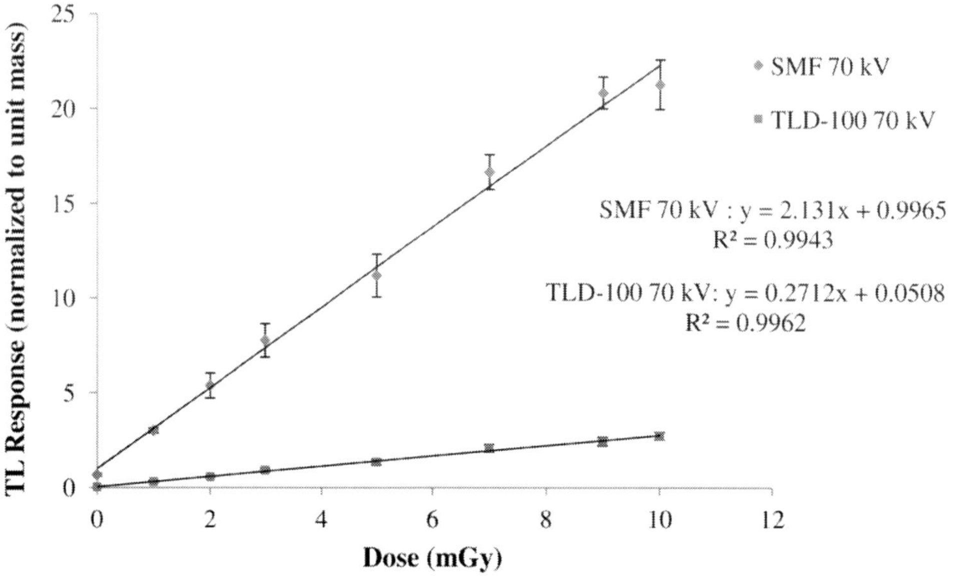

Figure 5. TL response against dose for SMF and TLD-100 (Rod).

The SMF shows significantly higher TL response is about 8 times more sensitive compared to that of TLD-100 (rod). The slope represents the TL sensitivity were 2.131 $mg^{-1}.Gy^{-1}$ and 0.2712 $mg^{-1}.Gy^{-1}$ for SMF and TLD-100 (rod), respectively. In terms of linear coefficient, the SMF shows $R^2 = 0.9943$ and $R^2 = 0.9962$ for TLD-100. As mentioned above, each single dose corresponds to the average of the three reading. The error bar for SMF and TLD-100 is in the range of 0.064-1.309 and 0.004-0.202, respectively. The error bar for TLD-100 also showed too small compare with SMF.

High energy x-ray photons will have a greater probability of penetrating the matter. In other words, the relatively low energy x-rays have a greater probability of being absorbed by the optical fibres [12, 13]. Therefore the lower the kV_p and mean energy, there will be greater photons absorption from the x-ray beam by the fibres. This will contribute to the better TL response of 30 kV compared to 70 kV from SMF and TLD-100.

4. Conclusion

In this paper, we have investigated the performance of SMF subjected to 30 and 70 kV x-rays irradiation and compared to TLD-100(rod). From our result, we found that the SMF have 10 and 8 times more sensitive than TLD-100 (rod) will point to its use in radiodiagnostic purpose.

Acknowledgements

The authors would like to thank The Ministry of Education (MOE) HIR grant (A000007-50001) for funding the fiber pulling project and Universiti Teknologi Malaysia for providing financial assistance through Research University Grant Scheme (RUGS), under Program Saintis Negara. Project number (10J37).

References

[1] Rivera T 2012 Thermoluminescence in medical dosimetry *Appl. Radiat. Isot.* **71** 30-34

[2] Hashim S, Al-Ahbabi S, Badley D A, Webb M, Jeynes C, Ramli A T and Wagian H 2009 The thermoluminescence response of doped SiO_2 optical fibres subjected to photon and electron irradiation *Appl. Radiat. Isot.* **67** 423-427

[3] Wagiran H, Hossain I, Bradley D A, Yaakob A N H and Ramli T 2012 Thermoluminescene response of photon and electron irradiated Ge- and Al- doped SiO_2 optical fibres *Chin. Phys. Lett.* **29** (2) 027802(1)-027802(3)

[4] Badita E, Stancu E, Scarlat F, Vancea C and Scarisoreanu A 2013 Study of optical fibe as new radiation dosimeter *Rom. Rep. Phys.* **65** (4) 1438-1442

[5] Saeed M A, Fauzia N A, Hossain I, Ramli A T and Tahir B A 2012 Thermoluminescence response of Germanium-doped optical fibers to x-ray irradiation *Chin. Phys. Lett.* **29** (7) 078701(1)-078701(3)

[6] Abdul Rahman A T, Hugtenburg R P, Abdul Sani S F, Alalawi A I M, Issa F, Thomas R, Barry M A, Nisbet A and Bradley D A 2012 An investigation of the thermoluminescence of Ge-doped SiO_2 optical fibres for application in interface radiation dosimetry *Appl. Radiat. Isot.* **70** 1436-41

[7] Bauk S, Alam M S and Alzoubi A S 2011 Precision of low-dose response of LiF:Mg, Ti dosimeters exposed to 80 kVp x-rays *J. Phys. Sci.* **22** (1) 125-130

[8] Girard S, Keurinck J, Ouerdane Y, Meunier J-P, Member IEEE, Member OSA and Boukenter A 2004 γ-rays and pulsed x-rays radiation response of Germanosilicate Single-Mode Optical Fibers: Influence of cladding codopants *J. Lightwave Technol.* **22** (8) 1915-1922

[9] ISO 1999 *The Int. Organization for Standardization. X and Gamma Reference Radiation for Calibrating Dosimeters and Dose Rate Meters and for Determining Their Response as a Function of Photon Energy* ISO 4037-3

[10] Pagonis V, Kitis G and Furetta C 2006 *Numerical and Practical Exercise in Thermoluminescence* (New York: Springer Science + Business Media)

[11] Portal G 1981 *Preparation and Properties of Principal TL Products, In: Applied Thermoluminescence Dosimetry*, ed M Oberhoffer and A Schermann (Bristol: Adam Hilger Ltd) pp 97-122

[12] Khan F M 2003 *The Physics of Radiation Therapy* (USA: Lippincott William & Wilkin)

[13] Knoll G F 2010 *Radiation Detection and Measurement* (USA: John Wiley & Sons)

Journal of Physics: Conference Series **546** (2014) 012019 doi:10.1088/1742-6596/546/1/012019

Cross coherence independent component analysis in resting and action states EEG discrimination

A Almurshedi and A K Ismail

Department of Physics, Faculty of Science, Universiti Teknologi Malaysia, 81310 Johor, Malaysia

E-mail: khamim@utm.my

Abstract Cross Coherence time frequency transform and independent component analysis (ICA) method were used to analyse the electroencephalogram (EEG) signals in resting and action states during open and close eyes conditions. From the topographical scalp distributions of delta, theta, alpha, and beta power spectrum can clearly discriminate between the signal when the eyes were open or closed, but it was difficult to distinguish between resting and action states when the eyes were closed. In open eyes condition, the frontal area (Fp1, Fp2) was activated (higher power) in delta and theta bands whilst occipital (O1, O2) and partial (P3, P4, Pz) area of brain was activated alpha band in closed eyes condition. The cross coherence method of time frequency analysis is capable of discrimination between rest and action brain signals in closed eyes condition.

1. Introduction

There are many differences in EEG signals between open and close eyes conditions. In different studies, different methods, such as power spectral analysis of EEG signals in open and closed eyes [1] and alpha activity in occipital lobe, are employed to analyse EEG signals. A study in 2007 [2] used Fourier transform to analyse EEG signals in open and closed eyes states, this study reported that the topographic changes were evident in all bands except for alpha band, and confirmed the use of mean alpha level as a measure of resting state arousal under open and closed eyes conditions. In a 2009 study, [3] researchers studied arousal vs. activation in children, their results confirmed the use of mean alpha level as a measure of resting state arousal under closed and open eyes conditions. Focal changes in the other bands reflect differences in activation; rather than the simple increase in arousal presented by alpha. Other techniques commonly utilized for EEG signals analysis include Fast Fourier transform, wavelet transformation dyadic filter bank, principal component analysis (PCA) and independent component analysis (ICA) [4, 5].

ICA has been used for EEG analysis in recent years [4]. It has been proven to be effective in removing the artifact from EEG signal [6]. Blink artifact is one of the very common noises in EEG signal. Cross Coherence is a statistical method that measures the degree of relationship, as a function of frequency, between two time series [7]. EEGLAB Matlab toolbox [8] is used to process the data and calculate ICA and cross coherence after discuss Brain Electrical Activity Mapping (BEAM) and power spectral analysis for two conditions open and close eyes in the resting and action states.

In this work, the traditional techniques Brain Electrical Activity Mapping (BEAM) were discussed. it can easily separate between open and closed eyes conditions, but cannot differentiate between resting and action states in closed eyes condition. For this purpose, ICA and cross coherence methods were used to study the relationship between the channels and components. The analysis of resting and action state of brain is very important to understand the behaviour of normal brain. Such study is helpful in monitoring of cognitive brain process including thinking, imagination, task difficulty and decision making and diagnosis of different brain diseases such as Depression, Migraine, Sleep Disorders, Learning Disabilities, Epilepsy and ADHD. This study provides evidences to the significance of ICA for removing the noise and cross coherence method for distinguishing between rest and action states of the brain in closed eyes condition.

2. Method and experiment

2.1. Participants

Twelve young students, all from Department of Physics in UTM, participated in this experiment. Their mean age was 25 (range 22–31) years old and all were right handed. Subjects were screened for neurological and physiological disorders, and no head injury. Some of them were wearing corrected sight vision lens during the experiment. The participants were voluntary and their written assent was obtained. They were informed about the protocol and the aim of this study. They were instructed to sit comfortably and to reduce the movement and blinking as possible; however they are asked to keep themselves attentive during the experiment. The experiment was conducted in the Instrumentation 4 Lab, the Department of Physics, Universiti Teknologi Malaysia.

2.2. EEG measurements

EEG data collected from 19-channel commercial KT88-2400 system (Digital EEG Topography). The impedance of each electrode was matched with the help of an LED indicator. Electrodes positions were distributed using 10-20 electrode system as suggested by [9]. The right and left hemispheres were referenced to the right and left earlobes (A1, A2) respectively and grounded to the forehead (A). The data was digitized at 200 Hz sampling frequency, and filtered using bandpass filter with 1-35 Hz frequency band. The routine continuous EEG data was recorded for each 3 min in two conditions, when the eyes are opened and closed, and in two states during resting and action state. The recorded data was saved on the hard disc for further offline processing. For each subject in each condition, the data is recorded from 19 electrode (Fp1, Fp2, F3, F4, F7, F8, Fz, T3, T4, C3, C4, Cz, T5, T6, P3, P4, Pz, O1, O2), power spectra were calculated using Fast Fourier Transforms. At each electrode, absolute power in the delta (0.8–3.8 Hz), theta (4–7.8 Hz), alpha (8–12.8 Hz) and beta (13–30 Hz) bands were calculated. ICA was performed using EEGLab Matlab toolbox after inserting of channel location. Furthermore, the time frequency analysis was applied to calculate cross coherence between active channels and components.

3. Independent component analysis

Independent Component Analysis is a method of decomposing a set of multivariate data into its underlying statistically independent components. Aapo Hyvärinen [10] rigorously defined ICA using the statistical latent variables model. Under this model, there is n random variables X_1, X_2, ..., X_n which are linear combinations of n random latent variables S_1, S_2, ..., S_n, as in equation (1). The latent variables are the sources S_i, which are statistically independent. They are called latent because they cannot directly be observed. They are also called as "independent components".

$$X = a_{\,1}S_1 + a_{\,2}S_2 + \ldots + a_{in}S_n \tag{1}$$

Both the independent components S_i and the mixing coefficients a_{ij} are unknown and must be estimated using the observed data X_i. The ICA latent variables model is better represented in matrix form [11]. If, $S = [S_1, S_2, S_3, \ldots, S_n]^T$ represents the original multivariate data that is transformed through some transformation matrix H producing X, as in equation (2), then ICA tries to identify an unmixing matrix W, as described in equation (3).

$$X = H\,S \tag{2}$$
$$W \approx H^{-1} \tag{3}$$

So, that the resulting matrix Y is:

$$Y = WX = W(HS) = S' \approx S \tag{4}$$

ICA demands that the original signals S_1, S_2, ... S_n, be at any time instant, statistically independent and the mixing of the sources be linear. In case of EEG signals, m-scalp electrodes picking up correlated signals from brain. EEG is a multivariate data interfered with muscle artifacts, such as eye blinks and head movements. Assuming that EEG data can be modelled as a collection of statistically independent brain signals [11].

4. Cross coherence

The amount of phase stability between two different time series can be measured, which is called coherence. Coherence combines the phase angle between two signals. Coherence is either one, when the phase difference between signals is constant, or zero, when the phase difference between signals is random. In certain cases, a constant phase angle difference appears between two different frequencies. This is termed as cross-frequency coherence or bi-spectral coherence [12, 13]. If the measures are within the same frequency band, then the terminology is basically coherence which assumes auto-frequency coherence. Coherence is a statistic of phase differences and yields a much finer measure of shared energy between mixtures of periodic signals. In fact, coherence is essential because the degree of relationship or coupling between any two living systems cannot be fully understood without the knowledge of its frequency structure over a relatively long period of time.

5. Result and Discussion

From the power spectrum in both states (i.e. resting and action), it has been noticed that the power spectrum was higher in the electrode position O1,O2, P3,P4, Pz. Based on BEAM, it is very difficult to differentiate between these two states since they are very close to each other in their frequency spectrum.

Table 1. Power spectrum from electrodes distribution of open, close, resting and action states.

Electrode Position	Open Eye (μv^2/Hz)	Close Eyes (μv^2/Hz)	
		Resting	Action
Fp1	65.7	7.3	9.0
Fp2	95.9	7.2	8.8
O1	10.6	51.0	53.1
O2	3.6	47.3	42.6
P3	14.0	46.0	42.7
P4	18.4	50.7	51.4
Pz	17.5	50.9	50.4

The power spectrum of seven selected electrodes (Fp1, Fp2, O1, O2, P3, P4, Pz) are shown in Table 1. These positions were selected due to its variances, as it shows the differences in the power frequency during open and close eyes. In open eyes condition, the highest power was at the frontal lobe (Fp1, Fp2) (95.9), but when the eyes were closed the power was relatively smaller, it was distributed between occipital (O1, O2) and parietal (P3, P4, Pz) the maximum was at occipital lobe (53.1) and these show good agreement with [14]. The activation of the brain lobe is based on the maximum frequency spectrum available at a specific electrode position.

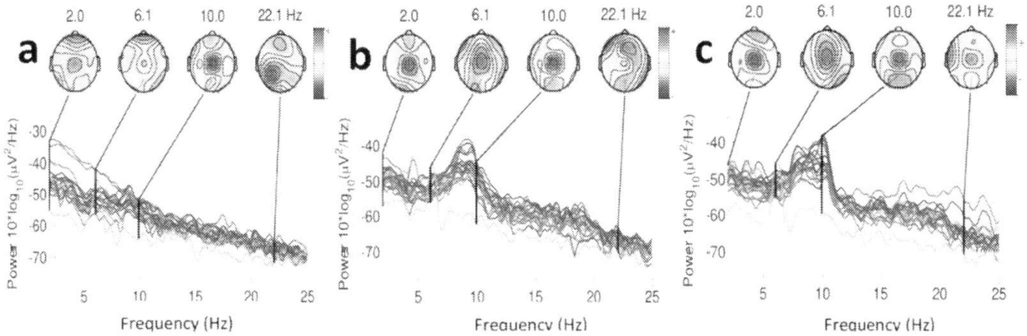

Figure 1. Channel power spectrum and topographical maps a) open eyes b) resting state c) action state.

The spectra of all electrodes, for open and close eyes conditions are presented in Figure 1. Clear differences can be observed between the spectra of open eyes (Figure 1a), close eyes rest state (Figure 1b) and close eyes action state (Figure 1c). As can be seen from Figure 1 (a), the frontal lobe is active with a higher power at low frequency and reduces gradually as the frequency increase. In Figure 1 (b,c), the higher power is moved to parietal and occipital lobes, while the prominent peak remained around 10 Hz. Comparing two conditions (open and close eyes), the power of delta, theta and alpha were intense in both conditions.

The topographical distributions of power and frequency bands are shown in Figure 2. The left columns (Figure 2a) present activity theta, delta, alpha and beta for open eyes condition. The middle columns (Figure 2b) depict brain activity in the resting state for close eyes condition, while the right set of columns (Figure 2c) shows the brain activity of action stat. In Figure 2, big differences for open eyes condition where the theta and delta power is much greater than for close eyes condition whereas the early alpha power is greater for the close eyes condition. The difference is almost double in both conditions. This result supported that Parieto-Occipital alpha waves during periods of eyes closed are the strongest EEG brain signals [15]. There was no noticeable marked difference in beta band. Based on Figure 2 it is very clear to distinguish between open and close eyes conditions. Since in open eyes condition, the frontal area is active with a high power of delta and theta bands whereas in close eyes conditions for both states, resting and action, alpha wave is active in partial-occipital area. These results are very similar to the open vs. closed eyes differences previously published [2].

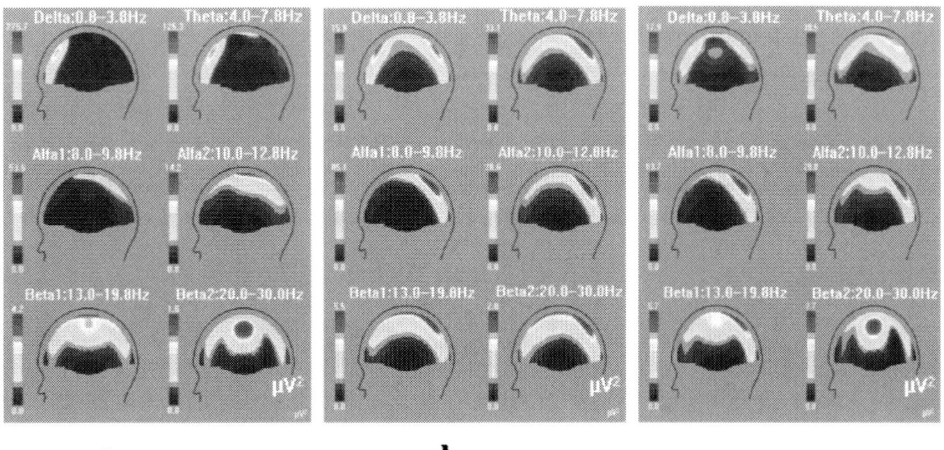

Figure 2. Brain electrical activity mapping (BEAM) for open eyes (a), close eyes resting state (b) and close eyes action state (c).

BEAM shows the topographical distribution and power spectrum frequency difference between open and close eyes. It also shows the activity of occipital and parietal lobes of brain in both resting and action states, but cannot easily differentiate between action and resting state of brain. For this reason, we tried to apply independent component analysis (ICA) and cross coherence between active channels and components.

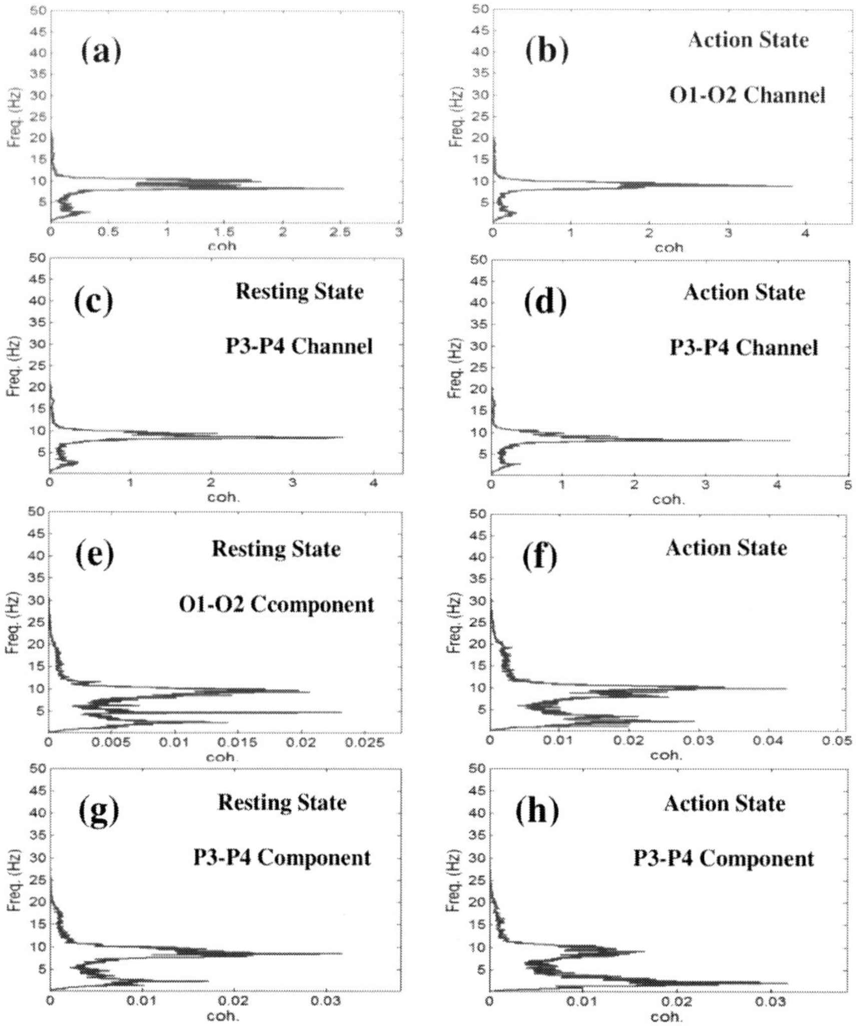

Figure 3. Cross coherence between channels and components right for resting state and left for the action state.

Figure 3 shows the result of cross coherence between EEG channels and components. In Figure 3, the cross coherence between channels and components plotted vs. frequency. The left column (Figure 3 a, c, e & g) for the resting state and the right column (Figure 3 b, d, f & h) for the action state. The first four graphs in Figure 3 are channel cross coherence between (O1-O2) and (P3-P4) in Figure (3a, 3b) and Figure (3c, 3d) respectively. Whereas the last four graphs show the component cross coherence of (O1-O2) and (P3-P4) in Figure (3e, 3f) and Figure (3g, 3h) respectively. The cross coherence between the channels observed with a sharp clear peak around 10 Hz. After applying ICA, the cross coherence between the components cover slower frequency area reaching 10 Hz, this happened because ICA removed the blink artifact and noise from the signals [6]. From Figure 3, there is more cross coherence between channels and components in action state than in resting state. This result indicates that there is more coherence between right and left hemisphere in action stare of brain than that in the resting state.

6. Conclusion

In this paper, the power spectrum brain topography was applied to analyse and distinguish EEG signal between close and open eyes conditions. Results confirmed the difference of power activation since delta and theta frequency band in the frontal brain lobe were active in the open eyes condition and early alpha wave in the parietal occipital brain lobe was active in close eyes condition. The ability and weakness of this method were discussed to differentiate between the resting and action states for close eyes condition. A method of using independent component analysis (ICA) and cross coherence time-frequency analysis was proposed to show the difference between these states. The proposed method show that the cross coherence between channels and components is always greater in the active state than resting state, i.e. there is stronger relation between both hemispheres in action state than that in the resting state. These results present the relationship between the channels and components and also show a great finding of power disruption in the brain for different conditions and states. The cross coherence method succeeds to distinguish between action and resting state of brain.

Acknowledgement

The authors would like to thank Department of Physics faculty of Science, Universiti Teknologi and Ministry of Higher Education for the financial supports. This research supported by Research University Grant, Vot Number: Q.J130000.2526.07H93.

References

[1] Ghiyasvand M, Guha S, Anand S and Deepak K 1995 14th Conf. of the Biomedical Engineering Society of India IEEE 3/61-3/62
[2] Barry R J, Clarke A R, Johnstone S J, Magee C A and Rushby J A 2007 *Clinical Neurophysiology* **118** 2765-2773
[3] Barry R J, Clarke A R, Johnstone S J and Brown C R 2009 *Clinical Neurophysiology* **120** 1806-1811
[4] Iriarte J, Urrestarazu E, Valencia M, Alegre M, Malanda A, Viteri C and Artieda J 2003 *Journal of clinical neurophysiology* **20** 249-257
[5] Akin M 2002 *Journal of Medical Systems* **26** 241-247
[6] Jung T-P, Humphries C, Lee T-W, Makeig S, McKeown M J, Iragui V and Sejnowski T J 1998 *Advances in neural information processing systems* 894-900
[7] Delorme A, Makeig S, Fabre-Thorpe M and Sejnowski T 2002 *Neurocomputing* **44–46** 1057-1064
[8] Delorme A and Makeig S 2004 *Journal of Neuroscience Methods* **134** 9-21
[9] Homan R W, Herman J and Purdy P 1987 *Electroencephalography and clinical neurophysiology* **66** 376-382
[10] Hyvarinen A 1999 *Neural computing surveys* **2** 94-128
[11] Mozaffar S and Petr D 2002 *Information Telecommunication and Technology Center, University of Kansas, Lawrence, KS, Tech. Rep. ITTC-FY2003-TR-03050-02*
[12] Schack B, Vath N, Petsche H, Geissler H-G and Möller E 2002 *International Journal of Psychophysiology* **44** 143-163
[13] Schack B, Klimesch W and Sauseng P 2005 *International Journal of Psychophysiology* **57** 105-114
[14] Li L, Xiao L and Chen L 2009 *Journal Of Electronic Science And Technology Of China* **7** 175-179
[15] Palva S and Palva J M 2007 *Trends in Neurosciences* **30** 150-158

Delivering key signals to the machine: seeking the electric signal that muscles emanate

A Y Bani Hashim[1], M N Maslan[1], R Izamshah[1] and I S Mohamad[2]

[1] Faculty of Manufacturing Engineering, Universiti Teknikal Malaysia Melaka, 76100 Durian Tunggal, Melaka, 76100, Malaysia
[2] Faculty of Mechanical Engineering, Universiti Teknikal Malaysia Melaka, 76100 Durian Tunggal, Melaka, 76100, Malaysia

E-mail: yusairi@utem.edu.my

Abstract. Due to the limitation of electric power generation in the human body, present human-machine interfaces have not been successful because of the nature of standard electronics circuit designs, which do not consider the specifications of signals that resulted from the skin. In general, the outcomes and applications of human-machine interfaces are limited to custom-designed subsystems, such as neuroprosthesis. We seek to model the bio dynamical of sub skin into equivalent mathematical definitions, descriptions, and theorems. Within the human skin, there are networks of nerves that permit the skin to function as a multi dimension transducer. We investigate the nature of structural skin. Apart from multiple networks of nerves, there are other segments within the skin such as minute muscles. We identify the segments that are active when there is an electromyography activity. When the nervous system is firing signals, the muscle is being stimulated. We evaluate the phenomena of biodynamic of the muscles that is concerned with the electromyography activity of the nervous system. In effect, we design a relationship between the human somatosensory and synthetic systems sensory as the union of a complete set of the new domain of the functional system. This classifies electromyogram waveforms linked to intent thought of an operator. The system will become the basis for delivering key signals to machine such that the machine is under operator's intent, hence slavery.

1. Introduction

Due to the limitation of electric power generation in the human body, present human-machine interfaces have not been successful because of the nature of standard electronics circuit designs, which do not consider the specifications of signals that resulted from the skin. In general, the outcomes and applications of human-machine interfaces are limited to custom-designed subsystems, such as neuroprosthesis.

People who apply prosthesis commonly claimed they felt a lack of feedback for the replacement limbs that requires great effort and concentration for them to execute a routine task. This is the effect of the absence of a direct neural interface that provides a comprehensive sensory feedback. The advances of modern science help more people those who used the prosthesis to replace their lost limbs due to birth defects, diseases or accidents. Intracortical micro stimulation percepts may be implemented to a somatosensory prosthesis where a much more detailed feedback could be obtained as compared to vibrotactile motors and sensory substitution devices that convey unrefined sensations [1].

2. Background

Stimulation of electric pulse that is localized in human skin is known as somatosensory evoked potential. It is a result of an electric response of the central nervous system to an electrical stimulation of peripheral nerve, for example, in the median nerve at the wrist or the posterior tibial nerve at the ankle [2]. Somatosensory input allows an upper-limb neuroprosthesis be connected with a brain-machine interface [3]. This brought out the somatosensory percepts for object manipulation process.

The nervous system mechanism can generate significant information towards electric physiological diagnosis and intraoperative monitoring [4,5]. In addition; micro stimulation can be used to activate localized populations of neural elements. However, predicting and subsequently controlling nervous responses to simultaneous current injection through multiple electrodes is a complex process. The computed electric field in the 3-D space surrounding the stimulating electrodes could overcome this drawback [6]. Furthermore, improving sensory and motor outcomes in chronic stroke patient and healthy subject through neuro rehabilitative treatments may be done by temporal functional deafferentation [7].

The body responds to the sub threshold stimulus when subjected to somatosensory stimulations to the tibialis anterior tendon while measuring the brain waves [8]. Therefore, a brain - computer interface may help individuals with disability to control assistive devices and reanimate paralytic limbs. The electrocorticography signals recorded from the sensorimotor cortex could be used for real-time device control in paralyzed individuals. Consequently, brain-computer interface introduces a direct link for transmitting communication between brain and artificial mechanism [9–11]. As a result; a controllable interface for assistive device could raise the quality of life for the disability [12,13].

One of the best approaches to interface with human body could be done through the invasive device [14]. For example, the combination of the hand-wrist splints, volitional and electrically stimulated muscle contraction could induce the corticospinal plasticity [15]. It is suggested that myoelectrically controlled functional electric stimulation clinically produces greater improvement similar to the physiotherapy session uses mock stimulation [16]. On the other hand, reading the data generated from an EMG-controlled stimulation system will have limitations such as the reliability of surface EMG (sEMG) recorded for a long period of time. In addition, the trustworthiness data of sEMG usage are the issue, when it comes to the sample size [15,17].

This paper explains the idea of developing a relationship between human and machine such that a human may instruct a machine to execute pre-thought activities via somatosensory activities generated in the brain. We are attempting natural interaction of a human to machine through a non-invasive approach, hoping to create the ability for a human operator to merge with manufacturing machinery in running a production process.

3. Approach

We seek to model the bio dynamical of sub skin into equivalent mathematical definitions, descriptions, and theorems. Within the human skin, there are networks of nerves that permit the skin to function as a multi dimension's transducer. We investigate the nature of structural skin. Apart from multiple networks of nerves, there are other segments within the skin such as minute muscles. We identify the segments that are active when there is an electromyography activity. When the nervous system is firing signals, the muscle is being stimulated.

3.1. Active segment

The experiment was noninvasive in nature where we performed sEMG. The probes or electrodes are attached onto the skin surface using standard patches to read voltage excitations on muscle stimulation. The reference coordinates to place the electrodes is explained in Figure 1. It shows a grid that estimates the location to place each of the electrodes. For example, a set of probes that is with red, blue, and black cables may be placed at (d,C)-red, (d,D)-blue, and (e,A)-black, respectively. The black-cabled electrode functions well when it is placed on a bony area. So, (e,A) is the best position on the grid.

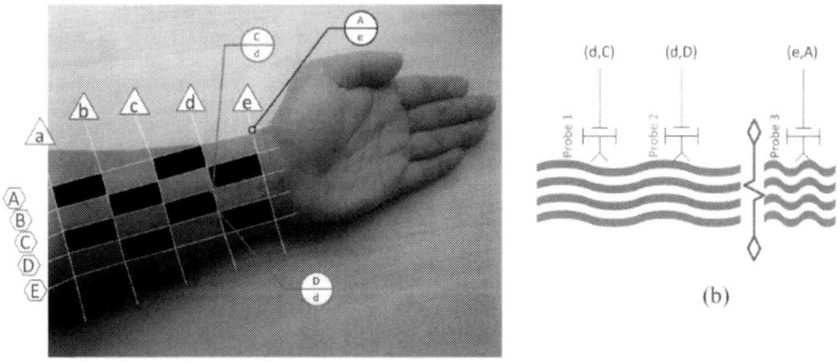

(a) (b)

Figure 1. (a) A grid that depicts the probable location where each of the probes may be attached. For example, there is a vertical axis $\langle a \rangle$ and a horizontal axis $[A]$ (b) The skin surfaces where all the nodes should be put. Probe 3, however, does not have the luxury of selecting a location. Its current location on the grid is fixed.

Table 1 lists the proposed locations for the electrodes. It explains the locations on the arm based on the anatomical landmarks [18]. The named locations, however, are shown based on the grid system mentioned in the preceding that corresponds to the landmarks.

Table 1. Active locations for electrode attachment.

№	Landmark	Estimated grid location
1	Brachioradialis	(a,A)
2	Flexor carpum radialis	(b,C)
3	Flexor carpum ulnaris	(c,E)
4	Smaller forearm extensors	inv(a,C) [a]

[a]At the back of the arm where the "inv" is a short form for "inverse".

3.1.1. Model 1. Let the grid, G has a horizontal axis with triangle scale notation for a through e, and a vertical axis with polygon scale notation for A through E such that $G=\{g \in G : g(\rho,\sigma)\}$ where $g(\rho,\sigma)$ is sub trees in the grid and ρ and σ represent vertex and edge, respectively that holds the information of the axes.

Let Ξ represents the abscissa where are definedts .the axis is defined as $\Xi=\{A,B,C,D,E\}$. Also, let Ψ represents the ordinate where are definedts .the axis is defined as $\Psi=\{a,b,c,d,e\}$. There exist a number of possible points that the electrodes may be placed with a number of possible combinations of coordinates. We have the following equation that explicitly defines the phenomenon.

$$\mathbf{X}=\{\chi \in X \subset \mathbf{G} : [\Psi \times \Xi] \mid \exists P \text{ may be fixed to some locations}\} \quad (1)$$

The sensor module has a set of three electrodes coded by colors, block A, and block B. Figure 2 depicts a generalized block diagram of the electronic system depicted in Figure 3. The probes 1, 2, and 3 are coded as red, blue, and black, respectively. Its present state is at (d,C), (d,D), and (e,A). Block A receives raw signals and amplifies them. Block B receives the amplified raw signals and rectifies them.

3.1.2. Model 2. Let P represents the probes or the electrodes. We have $P = \{p_1, p_2, p_3\}$. There are at most three electrodes. Let block A represents an amplifier circuit and block B a rectifier circuit. The generalized output function of the transducer system is:

$$\mathbf{O}(s) = \mathbf{P}(s) \cdot \mathbf{A}(s) \cdot \mathbf{B}(s) \tag{2}$$

Suppose there is a number of a combination of a set of probes to be localized within the grid, which is defined in equation (1). Each set of probes is represented by a sub tree and is related to a transducer system. Therefore, the root of the sub tree is the block B. We have for G:

$$\left.\begin{array}{l} \rho = \{A, B, p_1, p_2, p_3\} \\ \sigma = \{(A, B), (A, p_1), (A, p_2), (A, p_3)\} \end{array}\right\} \tag{3}$$

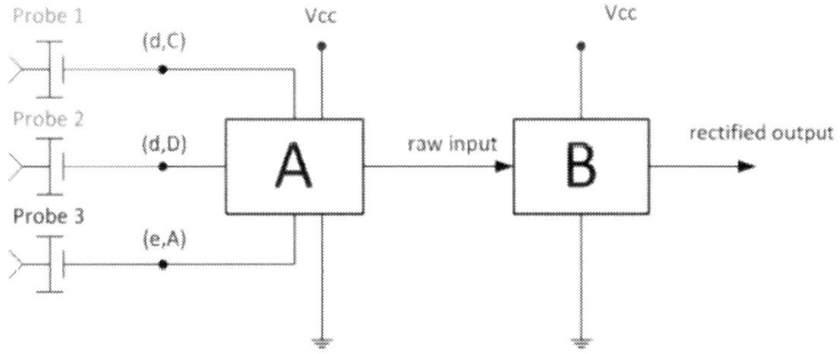

Figure 2. The transducer has three electrodes, circuit in block A, and circuit in block B. Block A receives signals from the muscles and produces raw outputs that are being amplified. Block B receives the amplified raw signals and produces rectified outputs.

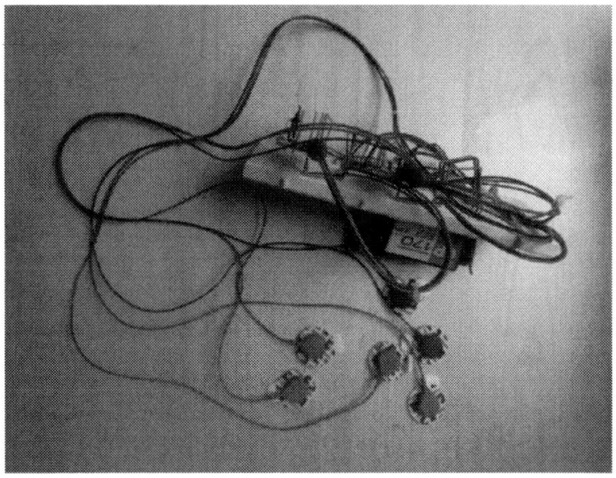

Figure 3. A transducer system prototype showing two sets of electrodes, circuit installed onto a breadboard powered by two units of 9-volt battery. This is a prototype-0.

3.1.3. Model 3. Figure 3 exhibits a pilot test on the transducer system where a set of electrodes was placed on a segment of the subject's left hand. The subject would have "intent to grab something." The muscle around that segment was stimulated, and an amount of potentials was read on the instrument.

Upon muscle stimulation, an amount of potential is transferred through the electrodes. These raw signals are being amplified and transfer to another process, which rectifies them. These useful signals are now ready to realize a human and a machine signal interface. The machine module will have a processor that accepts a certain signal protocol from the transducer system.

Let Π represents the machine processor protocol. We have the following equation that defines the human-machine signal interface protocol.

$$\Pi(s) = \begin{cases} 1, \forall \mathbf{O}(s) \text{reaches threshold} \\ 0, \text{otherwise} \end{cases} \tag{4}$$

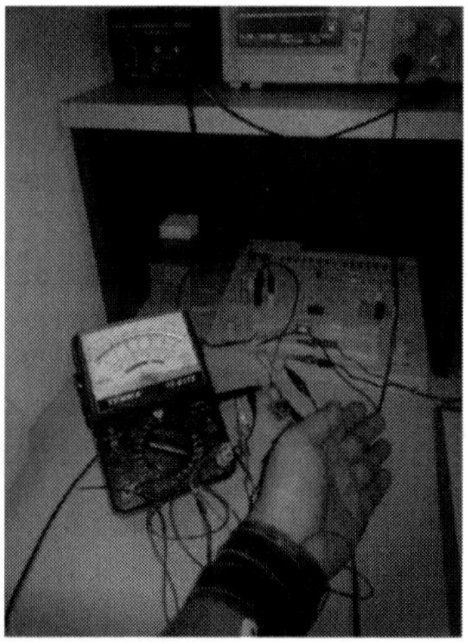

Figure 4. A pilot test of the transducer system was performed on the subject left hand. An oscilloscope and a multimeter were used to observe the evoked potentials. The instrument used were Tetronix TBS 1062 oscilloscope and Tenma 72-8170 multimeter.

4. Preliminary results

Ten male subjects participated in this study. The age ranges from 20-year-old to 27-year-old, all of whom have had no trauma. The experimental setup consisted of a Prototype-1 transducer system shown in Figure 5(b), a digital Parallax USB oscilloscope. The subjects were asked to supply the information that led to the body-mass index (BMI) shown in Table 2. The idea behind finding the BMI was to learn the degree of obesity that could have somehow affected the signal transmission from the skin to the electrodes due to fat layer thickness.

The forearm length was measured and assumed symmetry for both arms. Once registered, the subject was assigned an identification number. Based on the proposed landmarks on the forearm mentioned in Table 1, a number of grid location was identified and was suspected to be the locations where the electrodes should be placed, and the electromyograph could be read. There were three distinct muscles chosen, the flexor carpi radialis, the extensor digitorum superficialis, and the carpi radialis longus. The mean, median, and standard deviation for each of the grid locations are shown in Table 2, respectively. For the mean arm length of 24.90 cm, a grid location for flexor carpi radialis is somewhere at mean position (9.92 cm, 8.62 cm) where the origin is at the intersection of the vertical axis ⟨e⟩ and horizontal axis [e].

Table 2. Identified grid locations on forehand.

Subject	BMI	Arm length, L (cm)	Identified grid location $\{x, y: cm\}$		
			On flexor carpi radialis	**On extensor digitorum superficialis**	**On extensor carpi radialis longus**
1	19.72	27.0	10.8, 7.2	13.5, 2.3	13.5, 8.1
2	21.88	25.5	9.8, 8.0	12.3, 2.5	12.3, 9.0
3	23.29	26.0	10.4, 4.8	13.0, 2.5	13.0, 9.0
4	21.30	24.0	9.6, 7.2	12.2, 2.5	12.0, 8.1
5	27.70	28.5	11.4, 10.0	14.3, 3.1	14.3, 11.3
6	26.37	24.0	9.6, 9.6	12.0, 3.0	12.0, 10.8
7	25.00	23.0	9.2, 8.4	9.2, 2.6	11.5, 9.5
8	22.30	25.0	10.0, 10.2	12.5, 3.2	12.5, 11.5
9	26.00	23.0	9.2, 10.8	11.5, 3.4	13.5, 12.2
10	19.80	23.0	9.2, 10.0	11.5, 3.1	11.5, 11.3
Mean	-	24.90	9.92, 8.62	12.20, 2.82	12.61, 10.08
Median	-	24.50	9.70, 9.00	12.25, 2.80	12.40, 10.15
Std. deviation	-	1.77	0.71, 1.76	1.30, 0.36	0.89, 1.43

Figure 5 shows how the measurement was made on the forehand. This electrode positioning was based on the specific muscles identified earlier. The muscles involved in this procedure were the carpi radialis and brachioradialis. In this procedure, the second transducer prototype was used. Table 3 lists the settings, readouts, and oscilloscope's display for the three subjects chosen from the list seen in Table 2. The subjects were asked to move one or two of the fingers in a relax state while the measurement was made. In Table 3, subject-1 produced an output function corresponds to 70 mV, subject-2 140 mV, and subject-3 70 mV.

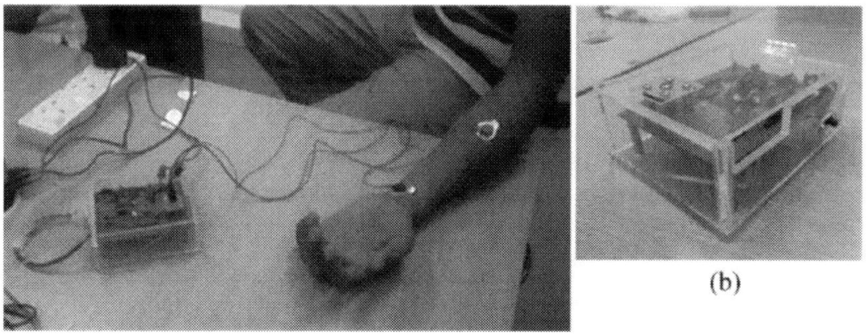

(a) (b)

Figure 5. (a) A test conducted where the electrodes were attached to the subject's left arm on carpi radialis (red electrode) and on brachioradialis (blue), while the black electrode functions as the frame ground. The electromyograph was measured using the transducer system and the Parallax USB oscilloscope. (b) This is the second prototype transducer system based on Prototype-0 with some modifications.

Journal of Physics: Conference Series **546** (2014) 012020 doi:10.1088/1742-6596/546/1/012020

Table 3. Readings obtained from the subjects on relax state.

Subject	Settings and readouts	Oscilloscope display (flexor carpi radialis)
1	Vertical: 1 V Horizontal: 2 ms Max. voltage: 150 mV Min voltage: -310 mV Pk-Pk voltage: 470 mV Mean voltage: -70 mV RMS voltage: 70 mV Period: 5.44 ms Frequency: 183 Hz $\therefore O(s) \mapsto 70mV$	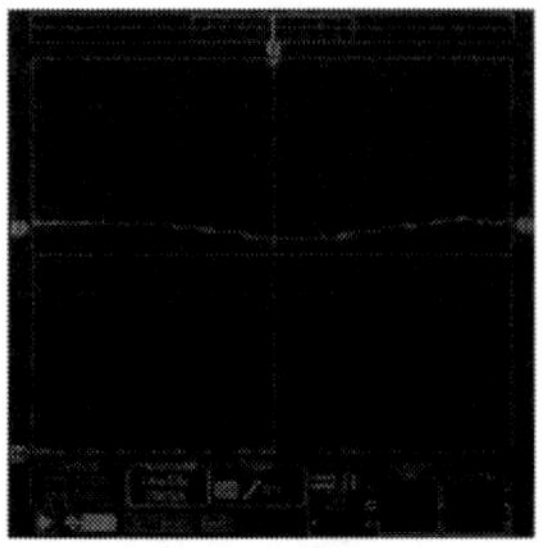
2	Vertical: 500 mV Horizontal: 2 ms Max. voltage: 200 mV Min voltage: -650 mV Pk-Pk voltage: 850 mV Mean voltage: -140 mV RMS voltage: 140 mV Period: 4.76 ms Frequency: 210 Hz $\therefore O(s) \mapsto 140mV$	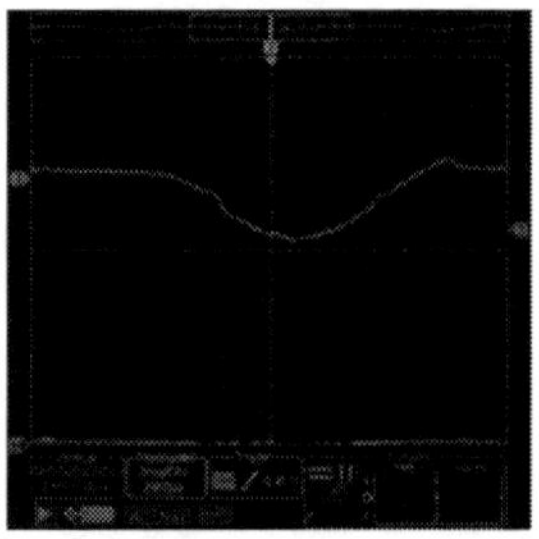
3	Vertical: 1 V Horizontal: 2 ms Max. voltage: 150 mV Min voltage: -470 mV Pk-Pk voltage: 620 mV Mean voltage: -70 mV RMS voltage: 70 mV Period: 2.92 ms Frequency: 342 Hz $\therefore O(s) \mapsto 70mV$	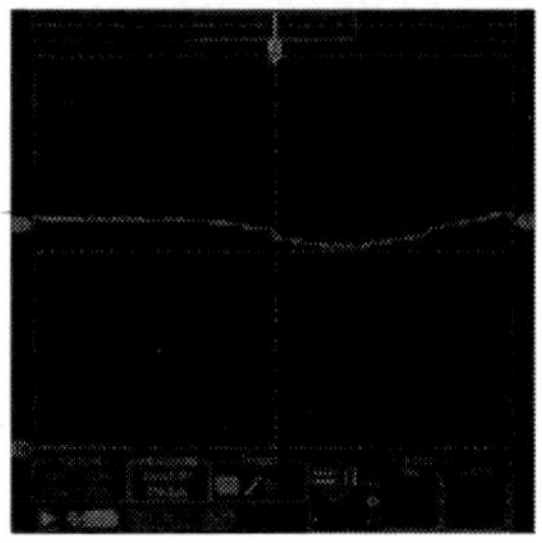

5. Conclusion

In result, we propose a relationship between the human and synthetic systems, sensory as the pairing of a complete circle of the new field of the functional scheme that classifies electromyogram waveforms linked to intent thought of an operator. The system will become the basis for delivering key signals to machine such that the machine is under operator's intent, hence machine slavery.

Acknowledgement

This research is supported by the Malaysian Ministry of Education, research grant identification number FRGS/2/2013/SG02/FKP/02/2/F00176. The experimental works were performed at the Robotics Laboratory at the Faculty of Manufacturing Engineering, Universiti Teknikal Malaysia Melaka, Malaysia, and were assisted by Nurulhuda Ahmad, Mohamad Farhan Mohamad Razif, Mohd Syahiran Md Amin, Abdullah Ikmar Kamaruddin, and Ahamad Zaki Mohammed Noor.

References

[1] Berg J A, Dammann J F, Tenore F V, Tabot G a, Boback J L, Manfredi L R, et al. 2013 Behavioral demonstration of a somatosensory neuroprosthesis. *IEEE Trans Neural Syst Rehabil Eng.* 2013 **21**(3) pp 500–7

[2] Zhang Z, Luk K D K, Hu Y 2010 Identification of detailed time-frequency components in somatosensory evoked potentials *IEEE Trans. Neural Syst. Rehabil. Eng.* **18**(3) pp 245–54

[3] Tabot G A, Dammann J F, Berg J A, Tenore F V, Boback J L, Vogelstein R J, et al. 2013 Restoring the sense of touch with a prosthetic hand through a brain interface *Proc. Natl. Acad. Sci. U S A* **110**(45) pp 18279–84

[4] Eberssole J S, Pedley T A 2003 *Current Practice of Clinical Electroencephalography*

[5] Niedermeyer E, da Silva F H L 2005 *Electroencephalography: Basic Principles, Clinical Applications, and Related Fields* (Philadelphia, Lippincott Williams & Wilkins)

[6] Choi J S, DiStasio M M, Brockmeier A J, Francis J T 2012 An electric field model for prediction of somatosensory (S1) cortical field potentials induced by ventral posterior lateral (VPL) thalamic microstimulation *IEEE Trans. Neural Syst. Rehabil. Eng.* **20**(2) pp 161–9

[7] Sens E, Knorr C, Preul C, Meissner W, Witte O W, Miltner W H R, et al. 2013 Differences in somatosensory and motor improvement during temporary functional deafferentation in stroke patients and healthy subjects *Behav. Brain Res.* **252C** pp 110–6

[8] So H-J, Oh H-Y, Kim S-H, Kim D-W 2013 A study on estimated sub-threshold intensities of somatosensory stimuli based on somatosensory evoked potentials of EEG *Third IEEE Int. Conf. Intell. Syst. Des. Eng. Appl.* pp 190–4

[9] Daly J J, Wolpaw J R 2008 Brain-computer interfaces in neurological rehabilitation *Lancet Neurol.***7**(11) pp 1032–43

[10] Donoghue J P 2008 Bridging the brain to the world: a perspective on neural interface systems *Neuron* **60**(3) pp 511–21

[11] Wang W, Collinger J L, Degenhart A D, Tyler-Kabara E C, Schwartz A B, Moran D W, et al. 2013 An electrocorticographic brain interface in an individual with tetraplegia *PLoS One* **8**(2) e55344

[12] Schalk G, Miller K J, Anderson N R, Wilson J A, Smyth M D, Ojemann J G, et al. 2008 Two-dimensional movement control using electrocorticographic signals in humans *J. Neural Eng.* **5**(1) pp 75–84

[13] Velliste M, Perel S, Spalding M C, Whitford A S, Schwartz A B 2008 Cortical control of a prosthetic arm for self-feeding *Nature* **453**(7198) pp 1098–101

[14] Tenore F, Vogelstein R 2011 Revolutionizing prosthetics: Devices for neural integration *Johns Hopkins APL Tech Dig* **30**(3): pp 230–239

[15] Fujiwara T, Kasashima Y, Honaga K, Muraoka Y, Tsuji T, Osu R, et al. 2009 Motor improvement and corticospinal modulation induced by hybrid assistive neuromuscular dynamic stimulation (HANDS) therapy in patients with chronic stroke *Neurorehabil. Neural Repair* **23**(2) pp 125–32

[16] Thorsen R, Cortesi M, Jonsdottir J, Carpinella I, Morelli D, Casiraghi A, et al. 2013 Myoelectrically driven functional electrical stimulation may increase motor recovery of upper limb in poststroke subjects: A randomized controlled pilot study *J. Rehabil. Res. Dev.* **50**(6) pp 785–94

[17] Yamaguchi T, Tanabe S, Muraoka Y, Imai S, Masakado Y, Hase K, et al. 2011 Effects of integrated volitional control electrical stimulation (IVES) on upper extremity function in chronic stroke *Keio J. Med.* **60**(3) pp 90–5

[18] Konrad P 2005 The ABC of EMG: Practical introduction to kinesiological electromyograph. *A practical introduction to kinesiological* pp 1–60

Correlation between near infrared spectroscopy and electrical techniques in measuring skin moisture content

M Mohamad, A R MSabbri, M Z MatJafri and A F Omar

School of Physics, Universiti Sains Malaysia, 11800 Penang, Malaysia

E-mail: mastura.mohamad02@gmail.com

Abstract. Near infrared (NIR) spectroscopy technique serves as an important tool for the measurement of moisture content of skin owing to the advantages it has over the other techniques. The purpose of the study is to develop a correlation between NIR spectrometer with electrical conventional techniques for skin moisture measurement. A non-invasive measurement of moisture content of skin was performed on different part of human face and hand under control environment (temperature 21 ± 1 °C, relative humidity 45 ± 5 %). Ten healthy volunteers age between 21-25 (male and female) participated in this study. The moisture content of skin was measured using DermaLab® USB Moisture Module, Scalar Moisture Checker and NIR spectroscopy (NIRQuest). Higher correlation was observed between NIRQuest and Dermalab moisture probe with a coefficient of determination (R^2) above 70 % for all the subjects. However, the value of R^2 between NIRQuest and Moisture Checker was observed to be lower with the R^2 values ranges from 51.6 to 94.4 %. The correlation of NIR spectroscopy technique successfully developed for measuring moisture content of the skin. The analysis of this correlation can help to establish novel instruments based on an optical system in clinical used especially in the dermatology field.

1. Introduction

The skin is composed of three structural layers with unique structure and function; stratum corneum (SC), epidermis, and the dermis [1]. The skin performs a wide variety of functions which the most important role to act as a physical barrier to the hostile external environment [2]. The outermost layer of epidermis, the stratum corneum (SC) is an important part in skin physiology and plays a vital role in the protection of humans [3] since it can act as the main protective barrier for skin to resists potential external mechanical, chemical and biological harmful effects [4]. The influence of the water content of SC on its mechanical properties has long been known. In 1952, Blank stated that water plays a preponderant role in the physical properties of the SC. The SC plays a vital role in maintaining life as a result of its water holding capacity and lipid content that functions in controlling the level of transcutaneous water loss [3][5]. It preserves the skin surface soft and smooth by binding the water even under very dry atmosphere [6]. Healthy and normal SC contains about 10-20 % water [7]. Failure of the SC to retain water induces dryness of the skin, producing a dry, scaly or rough skin surface.

Different non-invasive devices have been applied to evaluate the moisture content of the SC and other skin parameters for moisture determination as well as for detecting abnormalities in the skin [8]. The most popular technique is based on the measurement of electrical properties such as capacitance and alternating current [9]. Rosendal (1945), in the experiment showed that SC hydration was a dominating factor for its conductive properties [10]. Since the last decade, the developments of electrical devices for assessing the water content of skin are based on the electrical properties, such as measurement of resistance, capacitance, conductance and impedance. The basic technique relies on the behavior of the SC as a dielectric material with a sufficient amount of direct current resistance to act as the material between the plates of a capacitor [11]. However, this technique has some limitations such as lack of a direct correlation between skin conductivity and water content besides the flow of current in subject to variations in ion movements and re-orientation of protein dipole moments [8]. The electrical devices are influenced by external humidity and temperature that require them to be retained at a constant humidity and temperature [12]. Besides, the electrical conductance devices can be affected by the amount of electrolyte that skin contains as well as by the contact area of the surface of the probe on the skin [12-13]. Alternatively, several current researches have proposed the application of NIR spectroscopy in measuring skin moisture [14-18].

NIR spectroscopy is a type of vibrational spectroscopy located in the wavelength range of 750 to 2500 nm between the red band of visible light and the mid infrared (mid-IR) region. The NIR signal (spectrum) is a consequence of the light absorption caused by molecular vibrations (overtones and combinations of fundamental vibrations) of hydrogen bonds like C-H, N-H, O-H and –SH functional groups [19]. In the NIR region, the O-H bands of water are very intensive which exhibit five absorption maxima at 760, 970 nm (due to the second overtone of the O-H stretching band), 1190nm (the combination of the first overtone of the O-H stretching and the O-H bending band), 1450 nm (first overtone of the OH stretching), 1940 nm (due to combination of the O-H stretching band and the O-H bending band) [20-21]. Since 1983, a pioneering work conducted by Ellis and Bath on the analytical growth of the NIR spectra region has been carried out for the determination of water in gelatin by using its stronger absorption in the NIR region [22]. In 1980s, Norris et al. developed the first NIR spectroscopy for analysis of agriculture product [23]. Measuring oxygen saturation of hemoglobin is one of the most successful NIR spectroscopy applications in the field of biomedical engineering [9].

The NIR technology, from the last few decades, has accomplished numerous developments in many fields and has been regarded as the widely used analytical technique in various domains including the biomedical science analysis. This rapidly growing interest in the NIR spectroscopy lies in its benefits over other instrumental technique. It is because of the non-invasive and non-destructive character of this analysis; sample can be characterized by or without minimal sample preparation of demands. Another advantage is, as an analytical tools it is a very fast; provides spectra very quickly (one minute or less per sample) [21-22]. A number of detection surfaces and probes applied in measurement are occluded, therefore causes a false water build-up on the skin surface [24]. One of the significant advantages of NIR spectroscopy technique is for being a non-occlusive measurement [9].

The aim of this study is to develop a relationship (coefficient of determination, R^2) between NIR spectroscopy with conventional electrical techniques (as reference values) for skin moisture measurement to perform a comparative study on the moisture measurement accuracy (high value of R^2) when NIR technique is applied on 10 different individual. Besides, this study also perform comparative study on the measurement accuracy between NIR with two different electrical methods namely conductance (using Dermalab moisture probe) and capacitance (using Moisture Checker).

2. Materials and method

2.1. Subjects

A total of 10 healthy male and female volunteers (Malaysian) age between 21-25 years. They had neither a history nor the presence of any allergies and skin diseases. The subjects were acclimatized for at least 20 minutes in a temperature-controlled room with temperature 21 ± 1 °C and relative humidity 45 ± 5 % prior to the experiments.

2.2. Experimental design

A non-invasive measurement of skin hydration was performed on the left palm and right, forearm left and right, upper arm left and right, cheek left and right and forehead. The NIR spectroscopy technique (NIRQuest) was applied in the measurement of moisture content of skin to the nine different parts of human face and hand of each of the ten subjects (person). The conventional conductance (Dermalab moisture probe) and capacitance (Moisture Checker) techniques were used as reference values for the relative water content of the skin [21-22]. For each measurement, a 2 cm x 2 cm test site was applied with three replicated measurements and the average of every sample was calculated to represent the individual skin sample. The experiment was conducted in a closed room under controlled environmental conditions (temperature 21 ± 1 °C, relative humidity 45 ± 5 %).

2.3. Instrumentation

The NIRQuest (Ocean Optics, Inc. World Headquarters) spectrometer is a high performance InGaAs array detector as it covers the wavelength ranges between 900-2500 nm with the high sensitivity optical bench. The NIRQuest spectrometer is a low noise and low dark current and higher optical resolution 9.5 nm w/25 μm slit. The advantages of the features make it an ideal instrument for analyzing the moisture content of the skin. The spectrometer consists of the light source HL-2000 tungsten halogen lamp with spectral emission in the wavelength region of 360-2000 nm. Reflection probe used in this measurement is R600-7-VIS-125F that covers the wavelength range between 400-2500 nm. To calibrate the reflectance spectrum of halogen lamp, a reference spectrum was taken with the WS-1-SL, a white diffuse reflectance standard. During the calibration process, the reflectance probe was placed perpendicularly on the top of the surface white diffuse reflectance standard. The experimental diagram for calibration of NIR system is illustrated in Figure 1.

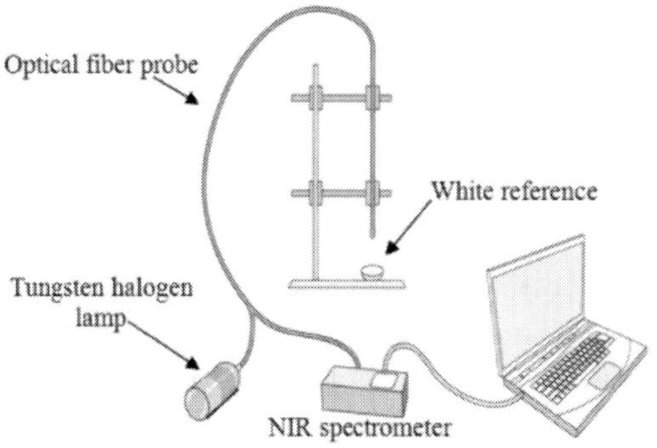

Figure 1. Schematic diagram for calibration of NIR spectroscopy technique.

The DermaLab® USB Moisture Module is manufactured by Cortex Technology, Hadsund, Denmark. The moisture probe constitutes an electrical impedance based device, operating at a single frequency of 100 kHz. It provides the information about the hydration state by measuring the conducting properties of the very upper layer of the skin, when subjected to an alternating voltage. This method is denoted as a conductance, measured in the unit of microSiemens (μS) ranging from 0 to 1999 μS with the resolution of 1 μS. The moisture probe is a flat-faced which provides a relatively large contact surface and is ideal for normal skin measurement. The probe consists of a 13 mm sensor diameter with spring loaded action organized as a concentric.

The Scalar Moisture Checker MY-808S (Scalar Corporation, Japan) is designed based on measuring the capacitance value of SC hydration. The Moisture Checker functions give positive correlations between the water content and the dielectric percentage. Therefore, by measuring the dielectric, the percentage of the moisture of the skin can be measured. The Moisture Checker has a higher measurement accuracy that provides a higher correlation with actual moisture content in the skin. Measured values are displayed in percentages (%).

3. Result and discussion

In the analysis of electrical conventional instruments, relative water content of the skin for all examined subjects obtained by using Dermalab moisture probe from the range 14 to 1120 μS and Moisture Checker from 16.5 to 39.7%.

The spectroscopic measurement performed on the 10 subjects at nine different parts of face and hand. The reflectance spectra for the nine different parts of face and hand on normal skin are shown in Figure 2. The absorption peaks of water in skin located approximately at wavelengths 970, 1190, 1450 and 1900 nm. The samples of skin have clearly demonstrated the prominent bands associated with water content of the skin. In the near infrared region with the wavelength range between 900 to 2500 nm, four absorption peaks of water for skin moisture could be observed. However, the weak absorption band around 970 nm shown and has more effect of light scattering due to the shortness of wavelength.

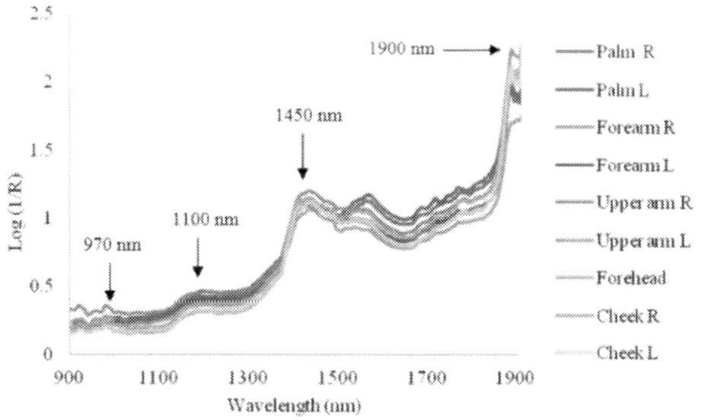

Figure 2. NIR Reflectance spectra of nine different parts of face and hand.

The analysis has been evaluated between the results obtained from different instrumental techniques and is focus to generate the correlation between NIR spectroscopy and electrical technique. NIR reflectance spectra and reference values were performed on the same sites and the correlation between the techniques were investigated. The value of R^2 of the ten subjects presents in Table 1.

Table 1. The experimental results conducted using a combination of four selected wavelengths (950-990 nm, 1100-1900 nm, 1400-1450 nm, 1870-1900 nm).

Subjects	Coefficient of determination, R^2 (%)	
	NIRQuest &Dermalab	NIRQuest & Moisture Checker
1	77.2	80.1
2	70.9	58.8
3	95.2	56.3
4	96.5	94.4
5	91.7	81.6
6	95.0	51.6
7	81.5	67.3
8	97.9	74.5
9	93.4	90.3
10	88.3	74.1
Average	88.8	72.9

From the experimental data summarized in Table 1, moisture content of skin determined by NIR spectroscopy technique correlated well with the electrical conventional techniques with the R^2 values range from 70.9 to 97.9 % (average of R^2 = 88.8 %) between NIRQuest and Dermalab moisture probe. Least accuracy has been observed for the correlation between NIRQuest and Moisture Checker with the R^2 values ranges from 51.6 to 94.4 % (average of R^2 = 72.9 %). The highest R^2 value has been generated efficiently produced linear relationship presented graphically in Figure 3 between NIRQuest and Dermalab moisture probe values (R^2 = 97.9 %), Figure 4 between NIRQuest and Moisture Checker values (R^2 = 94.4 %).

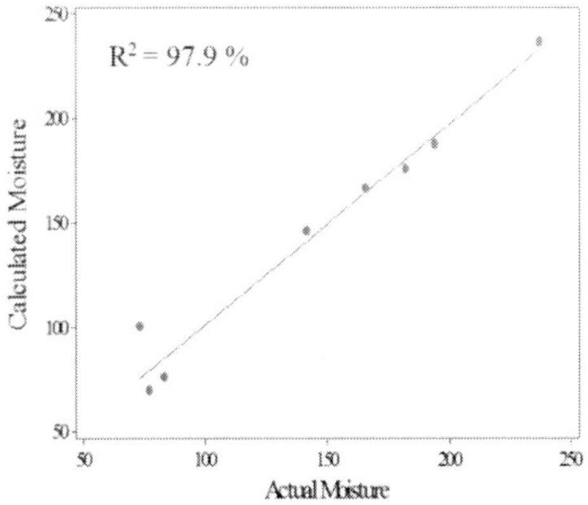

Figure 3. The scatter plot graph for the best correlation obtained between the values of NIRQuest spectrometer and Dermalab moisture probe (conductance).

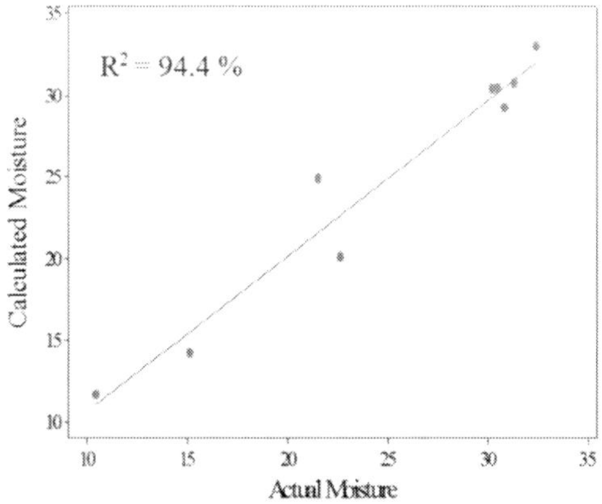

Figure 4. The scatter plot graph for the best correlation obtained between the values of NIRQuest spectrometer and Moisture Checker (capacitance).

The correlation between NIR spectroscopic and electrical conventional techniques successfully generated in measuring moisture content of the skin in addition to the advantages of NIR spectroscopy over the electrical techniques. The result from this study shows that measurement accuracy may vary from one individual to another. However, the deviation of accuracy is highly depended on the reference instrument. From this study, the higher consistency in measurement accuracy can be obtained when using conductance rather than capacitance technique as reference values.

The spectroscopic technique shows stability in the changes of temperature and relative humidity if compare to Dermalab moisture probe and Moisture Checker which is easily affected by environmental condition. The result obtained by using NIRQuest spectrometer is high in accuracy and precision due to low noise and dark current signal provide by the instruments. The most significant advantages is the non-occlusive measurement can be made by using spectroscopic technique, since the measurement can be performed without the interface of optical probe to be pressed with the skin surface.

4. Conclusion

This research has successfully generated correlation between NIR spectroscopy with conventional electrical techniques, Dermalab®USB Moisture Module and Scalar Moisture Checker. From the analysis of the entire research confirmed the ability of spectroscopic technique for practically used in quantifying moisture content of human skin. However, the accuracy of measurement may different when measured on different individual and when using different electrical methods as the reference values. In addition, the analysis of this correlation can assist to establish novel instruments based on an optical system in clinical application especially in the dermatology field.

References

[1] Stefaniak A B, Plessis J D, John S M, Eloff F, Agner T, Chou T C, Nixon R, Steiner M F C, Kudla I and Holness D L 2013 *Skin Res and Technol* **19** 59–68
[2] Geerligs M 2010 *Skin layer mechanics* Technische Universiteit Eindhoven
[3] Obata M and Tagami H 1989 *J. Invest. Dermatol.* **92** 854-59
[4] Pasumarty S M, Johnson S A, Watson S A and Adams M J 2011 *Tribol Lett* **44** 117–137
[5] Lodén M and Maibach H I 2012 Treatment of Dry Skin Syndrome: The Art and Science of Moisturizers: Springer
[6] Blank I H 1952 *J. Invest. Dermatol.* **18** 433-40
[7] Gabard B 2005 *Bioengineering of the skin: Water and Stratum Corneum* J Fluhr, P Elsner, E Berardesca and H I Maibach 2nd ed (Boca Raton: CRC Press) p 211-36
[8] Qassem M and Kyriacou P A 2012 Engineering in Medicine and Biology Society (EMBC), 34th Annual International Conference of the IEEE 6044-6047
[9] Arimoto H, Egawa M and Yamada Y 2005 *Skin Research and Technology* **11** 27-35
[10] Ollmar S and Nicander I 2005 *Bioengineering of the skin: Water and Stratum Corneum* J Fluhr, P Elsner, E Berardesca and H I Maibach 2nd ed (Boca Raton: CRC Press) p 335-49
[11] Edwards C and Marks R 2005 *Bioengineering of the skin: Water and Stratum Corneum* J Fluhr, P Elsner, E Berardesca and H I Maibach 2nd ed (Boca Raton: CRC Press) p 323-34
[12] Woo Y A, Ahn J W, Chun I K and Kim H J 2001 *Anal. Chem.* **73** 4964-71
[13] Suh E J, Woo Y A and Kim H J 2005 *Arch. Pharm. Res.* **28** 458-62
[14] Omar A F and MatJafri M Z 2012 Selected Topics on Optical Fibre Technology. M. Yasin, SW Harun, and H. Arof, editors. Intech/Rijeka, Croatia 229-246
[15] Qassem M and Kyriacou P A 2013a Engineering in Medicine and Biology Society (EMBC), 35th Annual International Conference of the IEEE 2392-2395
[16] Qassem M and Kyriacou P A 2013b *Skin Res and Technol* **19** (2) 137-144
[17] Arimoto H and Egawa M 2013 International Society for Optics and Photonics SPIE BiOS 85871T-85871T
[18] Arimoto H and Egawa M 2014 *Skin Res and Technol* (InPress)
[19] Luypaert J, Massart D and Vander Heyden Y 2007 *Talanta,* **72** 865-83
[20] Büning-Pfaue H 2003 *Food Chem.* **82** 107-15
[21] Reich G 2005 *Advanced Drug Delivery Reviews* **57** 1109-43
[22] Pasquini C 2003 *J. Braz. Chem. Soc.* **14,** 198-219
[23] Williams P and Norris K 2001 Near infrared technology in the agricultural and food industries, 2nd ed., *American Association of Cereal Chemistry*, Inc. St. Paul, MN, USA
[24] Walling P L and Dabney J M 1989 *J. Soc. Cosmet. Chem.* **40** 151-71

Post-dialysis urea concentration: comparison between one-compartment model and two-compartment model

N S Ahmad Tamrin and N Ibrahim

Department of Defence Science, Faculty of Defence Science and Technology, Universiti Pertahanan Nasional Malaysia, Kem Sungai Besi, 57000 Kuala Lumpur, Malaysia

E-mail: nabilahsolehah@gmail.com

Abstract. The reduction of the urea concentration in blood can be numerically projected by using one-compartment model and two-compartment model with no variation in body fluid. This study aims to compare the simulated values of post-dialysis urea concentration for both models with the clinical data obtained from the hospital. The clinical assessment of adequacy of a treatment is based on the value of Kt/V. Further, direct calculation using clinical data and one-compartment model are presented in the form of ratio. It is found that the ratios of post-dialysis urea concentration simulated using two-compartment model are higher compared to the ratios of post-dialysis urea concentration using one-compartment model. In addition, most values of post-dialysis urea concentration simulated using two-compartment model are much closer to the clinical data compared to values simulated using one-compartment model. Kt/V values calculated directly using clinical data are found to be higher than Kt/V values derived from one-compartment model.

1. Introduction

End stage kidney disease (ESKD) is the result of the deteriorating function of kidneys. It is associated with accumulation of metabolic waste products such as urea in the body fluid [1]. The function can be partially accomplished by dialysis treatment. Dialysis comprises of two main modalities; haemodialysis (HD) and peritoneal dialysis (PD). PD uses the patient's peritoneal membrane as the medium to exchange the toxins from blood into dialysate via osmotic forces while HD removes the toxins through artificial membrane housed in a dialyser via diffusion and ultrafiltration [2, 3]. Patient undergoes a haemodialysis treatment in 3-5 hours and repeated 2-3 times a week [4].

Urea kinetic modelling (UKM) had been introduced by National Cooperative Dialysis Study Group (NCDS). The introduction of UKM offered an opportunity to calculate the efficiency of a dialysis treatment [5]. In addition, it also explained the reduction in urea concentration during the dialysis treatment. UKM uses the concept of compartmental modelling which had been widely used in the analysis of physiological composition of body fluid. It visualised the compartment model as one or more compartments associated with the physiological spaces containing the body fluid. The assessment of dialysis dose is based on clinical parameter, Kt/V. K refers to the dialyser clearance, t refers to the duration of dialysis and V refers to the urea distribution volume at the end of the dialysis. A treatment is considered adequate if Kt/V achieved the minimum target of 1.2 [6].

This paper aims to compare the simulated values of post-dialysis urea concentration for both kinetic models with the clinical data obtained from the hospital and to present the difference between

Kt/V values derived from one-compartment model with the Kt/V calculated directly using the clinical data in term of ratio. The results are useful for the physicians to predict the post-dialysis urea concentration for any patient.

2. Methods

2.1. Subjects

A total of 32 stable haemodialysis patients from Hospital Sultanah Aminah, Johor, Malaysia were selected as the subjects of study. All patients had steadily undergone dialysis for more than 5 years period. The patients aged between 20 and 80 years old and body weight varies from 30–90 kg. The clinical data; pre-dialysis urea concentration, post-dialysis urea concentration, urea generation rate, body weight and dialyser clearance for 5 years period were recorded and used in the simulation study.

2.2. Urea kinetic modelling

The modelling consists of two kinetic models; one-compartment model and two-compartment model. One-compartment model visualised the body fluid as one compartment while two-compartment model divides the body fluid into two compartments; intracellular and extracellular compartments [7, 8]. In both kinetic models, the urea is assumed to be generated outside the body fluid. The schematic diagram of the models are shown in Figure 1and Figure 2.

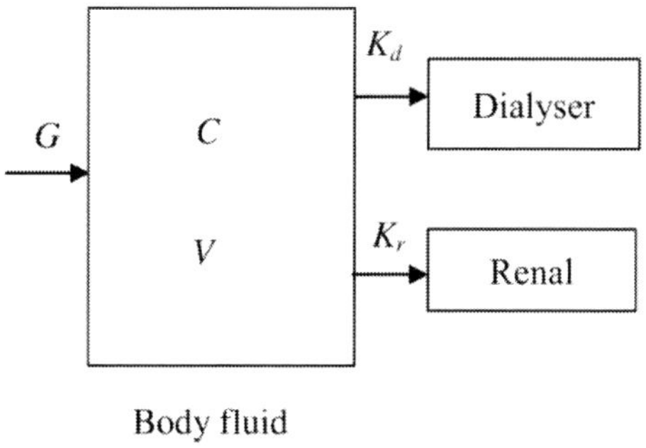

Figure 1. Schematic diagram of one-compartment model.

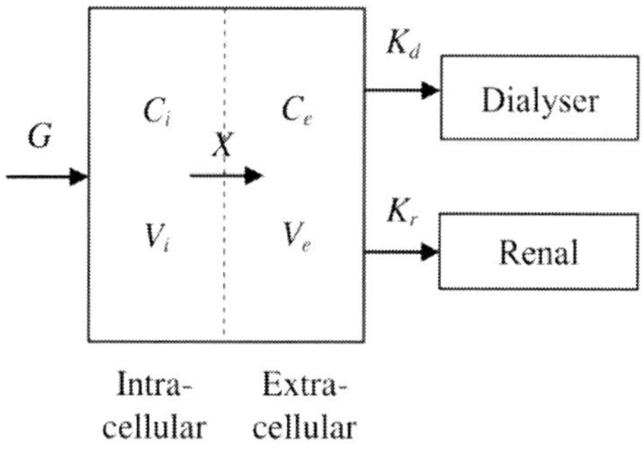

Figure 2. Schematic diagram of two-compartment model.

2.2.1. One-compartment model. Referring to Figure 1, based on law of mass conservation, the rate of change of mass of the body fluid can be represented by a differential equation.

$$\frac{\mathrm{d}(C(\mathrm{t}).V)}{\mathrm{dt}} = -K.C(\mathrm{t}) + G \quad (1)$$

where C(t) is the urea concentration (mg/ml) at time, t, V is the volume of solute distribution (ml) and G is solute generation rate (mg/min). K is the total clearance of urea (ml/min); $K = K_r$, between two dialysis and $K = K_d + K_r$, during dialysis. K_r and K_d are residual renal clearance and dialyser clearance, respectively.

The solution to equation (1) is given by as,

$$C(\mathrm{t}) = (C(0) - \frac{G}{K})\mathrm{e}^{-\frac{K.\mathrm{t}}{V}} + \frac{G}{K} \quad (2)$$

where C(t) and C(0) are urea concentration at any time, t and pre-dialysis urea concentration respectively.
Clinical parameter, Kt/V can be estimated using one-compartment model if pre- and post-dialysis urea concentration are known. It can be simplify as in equation (3). The ratio of urea generation rate and dialyser clearance are neglected in the calculation due to its small value compared to the value of pre-dialysis urea concentration.

$$\frac{K.t}{V} = \ln \frac{C(0)}{C(t)} \tag{3}$$

2.2.2. Two-compartment model. In this model, the volume, V of body fluid is divided into two compartments with volume V_i and V_e represent the intracellular and extracellular volume, respectively. Urea is cleared by the kidneys and/or the dialyser in the extracellular compartment. Referring to Figure 2, the rate of change of the body fluid mass is described by a set of two differential equations.

$$\begin{aligned} \frac{d(C_i(t).V_i)}{dt} &= -X(C_i(t) - C_e(t)) + G \\ \frac{d(C_e(t).V_e)}{dt} &= X(C_i(t) - C_e(t)) - K.C_e(t) \end{aligned} \tag{4}$$

where C_i(t) and C_e(t) is the urea concentration in the intracellular and extracellular compartment at time, t (mg/ml), respectively and X is the inter-compartmental clearance (ml/min).
The general solution for equation (4) is given as follows and solved using MATLAB software,

$$\begin{aligned} C_i(t) &= A_1 e^{\alpha_1 t} + B_1 e^{\alpha_2 t} + C_1 \\ C_e(t) &= A_2 e^{\alpha_1 t} + B_2 e^{\alpha_2 t} + C_2 \end{aligned} \tag{5}$$

where A_1, A_2, B_1, B_2, C_1 and C_2 are constants. α_1 and α_2 are time constants.

2.3. Simulations

In this simulation study, few assumptions were made. They are as follows: (i) the volume of urea distribution, V is assumed to be constant and V, V_i and V_e are estimated using equation (6) [9]. (ii) the residual renal clearance, K_r is approximately 0.001 which is equivalent to no urinary clearance and it is neglected in the calculation due to its small value compared to dialyser clearance [7]. (iii) inter-compartment clearance is assume to be 600ml/min [10] (iv) data such as body weight (BW), dialyser clearance (K_d), pre-dialysis urea concentration (C(0)), post-dialysis urea concentration (C (t=4hrs)), urea generation rate (G) and dialysis time (t_d) are from clinical data collected from the hospital. The simulated values were then compared with the clinical data. It is important to compare the models with the clinical data for validation purposes.

$$\begin{aligned} V(l) &= \frac{40}{70}.BW(kg) \\ V_i &= \frac{25}{40}.V \\ V_e &= \frac{15}{40}.V \end{aligned} \tag{6}$$

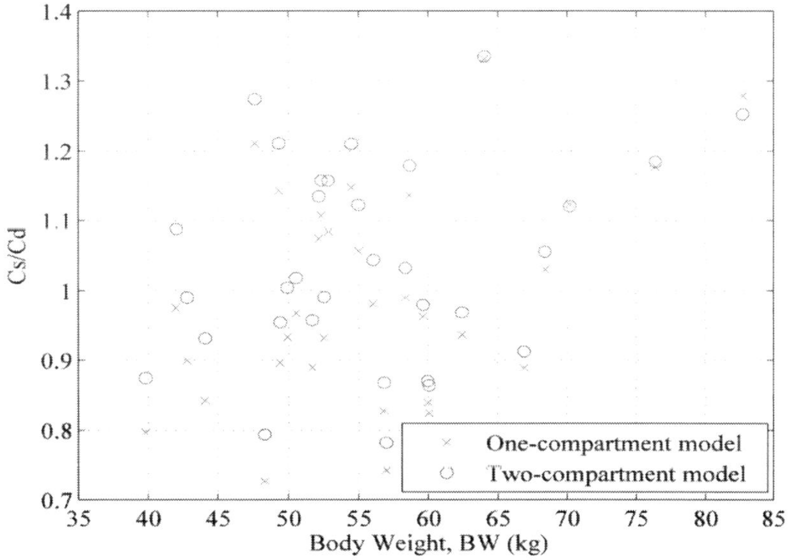

Figure 3. Graph of ratio of simulated data and clinical data for both kinetic models.

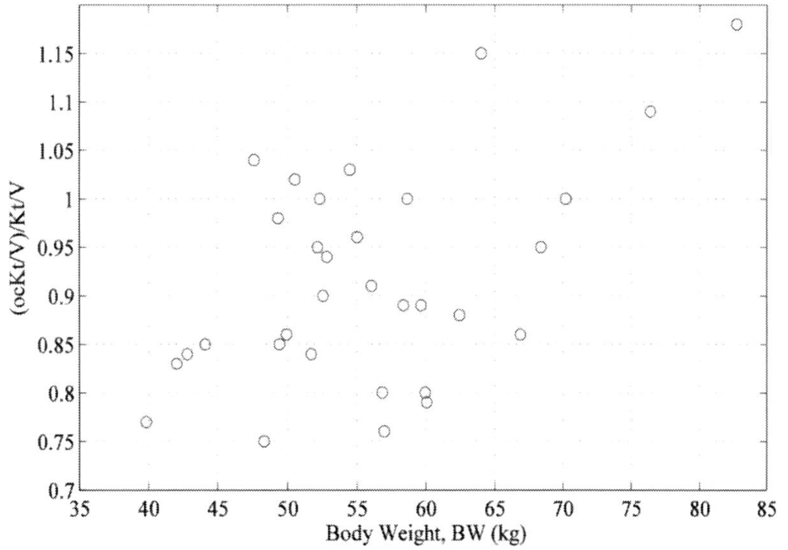

Figure 4. Graph of ratio of Kt/V values derived from one-compartment model (ocKt/V) and Kt/V values calculated directly using clinical data.

3. Results and Discussion

Figure 3 shows the ratio of simulated values with the clinical data for both models. Ratio of simulated and clinical data of post-dialysis urea concentration simulated using two-compartment model were found slightly higher compared to ratio of those simulated using one-compartment model. Most of the simulated values for one-compartment model underestimate (<1.0 ratio) the clinical data. It also shows that the simulated values of two-compartment model are closer to the clinical data compared to the values simulated using one-compartment model. Apart from that, the ratio of simulated values of post-dialysis urea concentration for both kinetic models are also found to be increase linearly with patients' body weight. The reduction in urea concentration are affected by the generation rate of urea. Larger patient have higher generation rate. Most of the simulated values had achieved the minimum target of urea reduction ratio (URR) $\geq 65\%$.

Figure 4 depicts the ratio of ocKt/V values derived from one compartment model and Kt/V calculated directly using the available clinical data. In the figure, most of ocKt/V values are found to be lower than the values of Kt/V. However, both calculation shows that they achieved the minimum target of Kt/V of at least 1.2. The SD of ocKt/V and Kt/V compared to the minimum target are 0.20 and 0.32. This shows that one-compartment model can be used to estimate the Kt/V value to determine the adequacy of a dialysis treatment.

4. Conclusion

The comparison of the values of post-dialysis urea concentration between one-compartment model and two-compartment model with the clinical data are presented. It is found that the ratios of post-dialysis urea concentration simulated using two-compartment model are higher compared to the ratios of post-dialysis urea concentration using one-compartment model. In addition, most values of post-dialysis urea concentration simulated using two-compartment model are much closer to the clinical data compared to values simulated using one-compartment model. Kt/V values calculated directly using clinical data are found to be higher than Kt/V values derived from one-compartment model.

Acknowledgements

The authors acknowledged the financial support from Ministry of Education Malaysia (MoE) in term of scholarship. The authors are grateful to Ministry of Health Malaysia (MoH) and Haemodialysis Unit, Hospital Sultanah Aminah, Johor for giving us the permission to collect the data.

References

[1] Walker W E, Hall D A, Sanfelippo M L and Swenson R S 1975 *Comput. Programs Biomed.* **5** 99
[2] Daugirdas J T, Blake P G and Ing T S 2007 *Handbook of Dialysis* vol 236 (Lippincott Williams & Wilkins)
[3] Levy J, Brown E, Daley C and Lawrence A 2009 *Oxford handbook of dialysis* (USA: Oxford University Press)
[4] Ing T S, Cheng Y L, Shek C C, Wong K M, Yang V L, Kjellstrand C M and Li C S 2000 *Hong Kong Journal of Nephrology* **2** 3
[5] Gotch F A and Sargent J A 1985 *Kidney Int.* **28** 526
[6] Waniewski J 2006 *J. Membr. Sci.* **274** 24
[7] Sanfelippo M L,Walker W E, Hall D A and Swenson R S 1978 *Comput. Programs Biomed.* **8** 44
[8] Burgelman M, Vanholder R, Fostier H and Ringoir S 1997 *Med. Eng. Phys.* **19** 69
[9] Lee C J and Chang Y L 1989 *Comput Methods Programs Biomed* **30** 33
[10] Clark W R, Leypoldt J K, Henderson L W, Mueller B A, Scott M K and Vonesh E F 1999 *J. Am. Soc. Nephrol.* **10** 601

AUTHOR INDEX